NUMERICAL TRIGONOMETRY

JOHN LOCKER
KENNETH F. KLOPFENSTEIN

COLORADO STATE UNIVERSITY

davies associates, publishers
Colorado

ii

davies associates
the academic publishing group
P.O. Box 440140
Aurora, CO 80044-0140

Library of Congress Cataloging-in-Publication Data

Locker, John
        Numerical trigonometry/John Locker, Kenneth F. Klopfenstein — 1st ed.
        Includes index
        ISBN 0-9630076-2-9
        1. Trigonometry   I. Klopfenstein, Kenneth F.   II. Title

    Library of Congress Catalog Card Number: 94-72314
                                                                        CIP

first edition
August, 1994

Editor: Elizabeth B. Davies
Editorial Assistant: Chad M. Mansfield, Colorado Springs, CO
Cover Design: Kenneth A. Martin, Martin Print Services, Boulder, CO
Printed in the United States

PREFACE

At Colorado State University precalculus mathematics is taught in a flexible, mastery-oriented instructional system called the Individualized Mathematics Program. This system is based on the belief that given proper organization and sequence of content, appropriate learning resources, adequate time, and sufficient positive feedback and encouragement, virtually every student can master the content of a course. The selection and organization of content and the style of presentation in this book reflects the mastery philosophy of the Individualized Mathematics Program.

The original version of these materials, which we wrote over a decade ago, was reproduced locally and used as one of several learning resources for students studying trigonometry in the Individualized Mathematics Program. We did not plan to become authors of a "real" textbook, but recent events caused us to reconsider. First, we realized how format and appearance influence the way students use textbooks and the credibility they assign them. A professionally produced volume is more used and more trusted than a sheaf of papers in a loose leaf binder. Second, the University of Colorado at Boulder began teaching precalculus mathematics, including the material in this book, by means of an instructional system similar to the Individualized Mathematics Program. The resulting increase in demand made local reproduction of the materials inconvenient for everyone involved. The preliminary edition of *Numerical and Analytic Trigonometry* was the first step toward transforming the notes, which served students well for many years, into a professionally produced text. This edition is the next step in that procedure. As we continue developing this text, we invite and welcome comments and suggestions from readers.

Carefully formulated instructional objectives serve as the fundamental organizational unit in *Numerical Trigonometry*. Each objective is prominently displayed and is followed immediately by focused discussion, summary, examples and practice problems. Various students will wish to use the materials following the objectives differently. Students who work from the general to the specific will want to digest the discussion, verify their understanding by comparing with the summary, work through the examples and, finally, do some practice problems. Other students with a different learning style may prefer to begin with concrete examples and practice problems, then read the summary to verify the insights derived from the examples, and finally, read the discussion to synthesize the information they have gained. Prominent headings invite each reader to use this material in the personally most efficient way.

Units of material are of modest length to provide manageable learning tasks and opportunities for frequent positive feedback and encouragement. In the Individualized Mathematics Program, at the end of each unit, students are required to demonstrate their competence on a short examination before proceeding to the next unit. Items on these unit exams are all derived from the objectives. Objectives are, in effect, templates for examination items. Sample unit exams are included at the end of each unit.

The discussions, summaries, examples, practice problems and sample examinations all focus on the stated objectives. We hope those readers who are learning trigonometry as a tool for solving triangles and those who are learning about the trigonometric functions as background for calculus will both find that these features make learning easier.

As we developed these materials, we benefited greatly from the comments of many students who learned from the original notes, from graduate and undergraduate course assistants who lectured and tutored from them, and from faculty colleagues who taught from them. We are especially indebted to Robert E. Gaines, our department head, for providing release time and summer support for us to write and develop the original notes, to Kristy Lahnert for typing the early versions of these materials, to our colleague Ervin Deal for proofreading original versions of the notes, and to Duane Kouba for working all of the practice problems for the manuscript.

We extend special thanks to Davies Associates, Publishers for their creative design, skillful and gentle editing, desk top publishing wizardry, and for their commitment to producing affordable, quality learning resources tailored to unique instructional programs.

John Locker
Kenneth F. Klopfenstein
Colorado State University
Fort Collins, Colorado
August, 1994

PUBLISHER'S ACKNOWLEDGMENTS

We would like to thank the authors, John Locker and Kenneth Klopfenstein, for their focus on mathematics as a *human* endeavor. We enjoy this working relationship and benefit from the learning environment. We are grateful for the authors' willingness to commit the time and resources to this project in order to make it successful. The student who chooses to use this textbook as a learning tool will be the best measure of the authors' success and ours. We welcome your comments.

Additionally, we wish to thank Kenneth A. Martin, of Martin Printing Services in Boulder, Colorado, for his cover design and artwork. Chad M. Mansfield, of Colorado Springs, Colorado, and formerly a student at Colorado State University, provided editorial assistance which helped to correct errors found in the preliminary edition of this book. We are grateful to everyone with whom we work for their humor and support. We are also indebted to Euclid for his contributions to this book.

This book was produced using software programs from Adobe Systems Inc. (Mountain View, California), Aldus Corporation (Seattle, Washington), Design Science, Inc. (Long Beach, California), and Microsoft Corporation (Redmond, Washington).

James Keith and Elizabeth Baker Davies
davies associates
the academic publishing group
August, 1994

CONTENTS

NUMERICAL TRIGONOMETRY

Other books by John Locker and Kenneth F. Klopfenstein:

Analytic Trigonometry

Other books by Kenneth F. Klopfenstein:

Exponential and Logarithmic Functions

NUMERICAL TRIGONOMETRY

FORMULAS FROM NUMERICAL TRIGONOMETRY

THE TRIGONOMETRIC FUNCTIONS:

$$\sin \theta = \frac{y}{r} = \frac{\text{side opposite } \theta}{\text{hypotenuse}}$$

$$\cos \theta = \frac{x}{r} = \frac{\text{side adjacent to } \theta}{\text{hypotenuse}}$$

$$\tan \theta = \frac{y}{x} = \frac{\text{side opposite } \theta}{\text{side adjacent to } \theta}$$

$$\csc \theta = \frac{1}{\sin \theta} = \frac{r}{y} = \frac{\text{hypotenuse}}{\text{side opposite } \theta}$$

$$\sec \theta = \frac{1}{\cos \theta} = \frac{r}{x} = \frac{\text{hypotenuse}}{\text{side adjacent to } \theta}$$

$$\cot \theta = \frac{1}{\tan \theta} = \frac{x}{y} = \frac{\text{side adjacent to } \theta}{\text{side opposite } \theta}$$

SPECIAL RIGHT TRIANGLES:

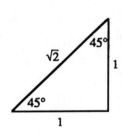

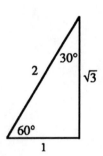

THE LAW OF SINES:

$$\frac{a}{\sin A} = \frac{b}{\sin B} = \frac{c}{\sin C}$$

THE LAW OF COSINES:

$$a^2 = b^2 + c^2 - 2bc \cos A$$

$$b^2 = a^2 + c^2 - 2ac \cos B$$

$$c^2 = a^2 + b^2 - 2ab \cos C$$

GRAPHS OF $y = A \sin B\theta$ AND $y = A \cos B\theta$:

$$A = \text{amplitude} \qquad\qquad \text{period} = \frac{2\pi}{B}$$

UNIT 1

ANGLES AND TRIGONOMETRIC FUNCTIONS

Introduction

The word "trigonometry" is a combination of the Greek words *trigon* (triangle) and *metron* (measure). Literally "trigonometry" means the measuring of triangles. In modern usage trigonometry has a broader meaning that includes the study of the relationships among the sides and angles of triangles (numerical trigonometry) and of the properties of the trigonometric functions (analytic trigonometry).

This Unit introduces basic concepts and terminology needed to study triangles. In particular, angles, degree and radian measure, and the trigonometric functions will be defined. Some elementary properties, such as the relationship between degrees and radians, will be studied and the signs of the trigonometric function in the various quadrants will be identified and examined.

UNIT 1
ANGLES AND TRIGONOMETRIC FUNCTIONS

Objective 1.1

(a) Given the degree measure of an angle,
- draw the angle in standard position and
- classify the angle according to its quadrant or as a quadrantal angle.

(b) Given a drawing of an angle in standard position,
- estimate its degree measure and
- classify the angle according to its quadrant or as a quadrantal angle.

Objective 1.2

Given the degree measure of an angle, find its radian measure. Conversely, given the radian measure of an angle, find its degree measure.

Objective 1.3

Given a point on the terminal side of an angle in standard position, evaluate the trigonometric functions for the angle.

Objective 1.4

Given information which determines the quadrant of an angle and given the value of one trigonometric function of that angle, find the values of the six trigonometric functions of that angle.

ANGLES AND TRIGONOMETRIC FUNCTIONS

Objective 1.1

    (a) Given the degree measure of an angle,
- **draw the angle in standard position and**
- **classify the angle according to its quadrant or as a quadrantal angle.**

    (b) Given a drawing of an angle in standard position,
- **estimate its degree measure and**
- **classify the angle according to its quadrant or as a quadrantal angle.**

Discussion

In trigonometry an **angle** is formed by the rotation of a ray about its endpoint. The endpoint is called the **vertex** of the angle. The beginning position of the ray is called the **initial side** of the angle and its final position after the rotation is called the **terminal side** of the angle.

Figure 1.1 shows angle POQ formed by rotating the ray OP through $\frac{1}{8}$ of a revolution to position OQ.

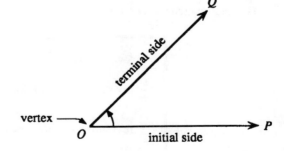

Figure 1.1

An angle formed by rotating a ray counterclockwise about its endpoint is called a **positive** angle. An angle formed by a clockwise rotation is called a **negative** angle. An example of each is shown in Figures 1.2 (a) and 1.2 (b). The first of these angles is labeled θ (*theta*), and the second is labeled ϕ (*phi*). Greek letters are often used as names for angles.

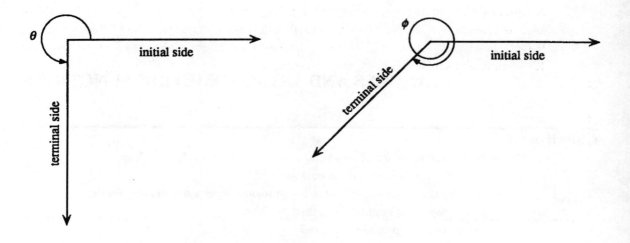

Figure 1.2 (a) A **positive** angle of $\frac{3}{4}$ revolution counterclockwise

Figure 1.2 (b) A **negative** angle of $1\frac{3}{8}$ revolutions clockwise

The most common unit for measuring angles is the **degree**. The angle formed by one full counterclockwise revolution is assigned a measure of 360 degrees, denoted 360°. Thus, **one degree,** denoted 1°, corresponds to $\frac{1}{360}$ of one full counterclockwise revolution. One full clockwise revolution is assigned a measure of −360°.

The degree measures of some standard positive angles are:

$$0 \text{ revolutions} = 0°,$$
$$\frac{1}{4} \text{ revolution counterclockwise} = \frac{1}{4} \cdot 360° = 90°,$$
$$\frac{1}{2} \text{ revolution counterclockwise} = \frac{1}{2} \cdot 360° = 180°,$$
$$\frac{3}{4} \text{ revolution counterclockwise} = \frac{3}{4} \cdot 360° = 270°,$$
$$2 \text{ revolutions counterclockwise} = 2 \cdot 360° = 720°.$$

Table 1.1

A **right angle** is a 90° angle. An **acute angle** is an angle between 0° and 90° (exclusive), an **obtuse angle** is an angle between 90° and 180° (exclusive). These are the angles which occur in triangles. In particular, a **right triangle** is a triangle that contains a right angle. Figures 1.3 (a), 1.3 (b), and 1.3 (c) show an example of each type of these angles.

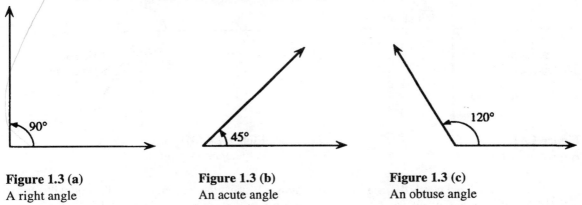

Figure 1.3 (a)
A right angle

Figure 1.3 (b)
An acute angle

Figure 1.3 (c)
An obtuse angle

We say that an angle is in **standard position** when its vertex is at the origin of a rectangular coordinate system and its initial side lies along the positive *x*-axis. The angles in Figures 1.4 (a), 1.4 (b), 1.4 (c), and 1.4 (d) are all in standard position. In addition, an angle in standard position is classified according to where its terminal side lies.

- An angle is a **quadrant I, quadrant II, quadrant III,** or **quadrant IV** angle when its terminal side lies in quadrants I, II, III, or IV, respectively. Figures 1.4 (a) - 1.4 (d) show examples of such angles.
- An angle is a **quadrantal angle** when its terminal side lies along the *x*-axis or *y*-axis. Angles of 0°, 90°, 180°, and 270° are examples of quadrantal angles.

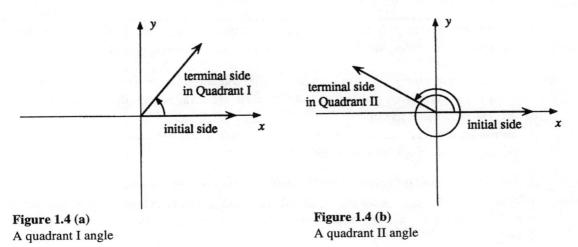

Figure 1.4 (a)
A quadrant I angle

Figure 1.4 (b)
A quadrant II angle

8

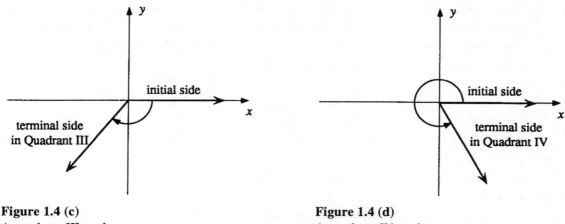

Figure 1.4 (c)
A quadrant III angle

Figure 1.4 (d)
A quadrant IV angle

The specialized words defined in this section will be used throughout this book. You will need to understand these terms in order to understand the book. Reading the Examples and doing the Practice Problems which follow will help you gain command of unfamiliar technical terms.

Summary

1. An **angle** is formed by rotating a ray about its endpoint. The endpoint about which the ray is rotated is called the **vertex** of the angle. The beginning position of the ray is called the **initial side** of the angle. The final position of the ray is called the **terminal side** of the angle. **Positive** angles are formed by rotating a ray counterclockwise; **negative** angles by rotating a ray clockwise; a **zero** angle by not rotating the ray at all.

2. The measure of an angle formed by rotating a ray through exactly one revolution counterclockwise is 360 degrees. Thus, **one degree** (1°) is $\frac{1}{360}$ th of a revolution counterclockwise.

3. A **right angle** measures 90°. Any triangle which includes a right angle is called a **right triangle**. An **acute angle** has measure between 0° and 90° (exclusive). Any triangle in which all angles are acute is called an **acute triangle**.
An **obtuse angle** has measure between 90° and 180° (exclusive). Any triangle which includes an obtuse angle is called an **obtuse triangle**.

4. An angle is in **standard position** when it is positioned in a rectangular coordinate system in such a way that the vertex of the angle coincides with the origin of the coordinate system and the initial side of the angle coincides with the positive x-axis of the coordinate system.

5. To draw an angle of degree measure $D > 0$ in standard position in a coordinate system, locate its vertex at the origin and its initial side along the positive x-axis. Then, draw its terminal side in the position reached by rotating the initial side counterclockwise about the vertex $\dfrac{D}{360}$ revolutions. Indicate the number of revolutions in the angle by drawing a spiral counterclockwise from the initial side to the terminal side of the angle.

To draw an angle of degree measure $D < 0$ in standard position in a coordinate system, locate its vertex at the origin and its initial side along the positive x-axis. Then draw its terminal side in the position reached by rotating the initial side clockwise about the vertex $\dfrac{|D|}{360}$ revolutions. Indicate the number of revolutions in the angle by drawing a spiral clockwise from the initial side to the terminal side of the angle.

Draw an angle of degree measure $D = 0$ in standard position in a coordinate system by locating its vertex at the origin and both its initial and its terminal sides along the positive x-axis.

6. To estimate the degree measure of an angle drawn in standard position, estimate how many revolutions (and fractions of revolutions) the initial side was rotated to reach the terminal side of the angle. Count counterclockwise rotations positively and clockwise rotations negatively. Multiply the estimated number of revolutions by $360°$ to obtain an estimate for the degree measure of the angle.

7. Angles in standard position are classified according to the location of their terminal side. An angle is called a **quadrant I, quadrant II, quadrant III,** or **quadrant IV** angle when the terminal side of the angle placed in standard position lies in **quadrant I, quadrant II, quadrant III,** or **quadrant IV,** respectively. An angle is called a **quadrantal angle** when the terminal side of the angle placed in standard position lies along the x-axis or y-axis.

Authors' Note

An angle is formed by turning or bending one side of the angle away from or toward the other side of the angle. The word angle comes from the Latin *angulus*, which means "a little bending." Interestingly, the Indo-European root "*ang-*" or "*ank-*" of the word *angulus* can be found in various English words like ankle, anchor and anglican, as well as the word English.

The angular-shaped island region found in Europe was originally called "Angul," later England. In the 5th century, the Teutonic people who settled there were called Angles, from *Angul-cynn* and *Angul-péod* (a race or a people from Angul).

Examples

Example 1. Sketch a 120° angle in standard position and classify it by quadrant or as a quadrantal angle.

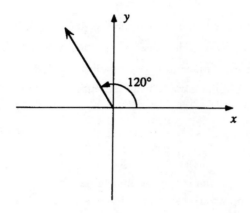

Solution 1.

Since 360° is one full revolution, the positive angle 120° is formed by

$$\frac{120}{360} = \frac{1}{3}$$

of one revolution **counterclockwise** from the positive x-axis.

The terminal side of the angle lies in quadrant II, so 120° is a quadrant II angle.

Figure 1.5

Example 2. Sketch a −45° angle in standard position and classify it by quadrant or as a quadrantal angle.

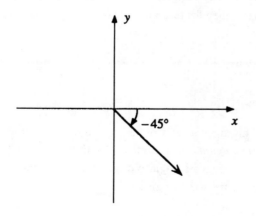

Solution 2.

A negative angle is formed by **clockwise** rotation. Since 360° is one full revolution, the angle −45° is formed by

$$\frac{45}{360} = \frac{1}{8}$$

of one revolution **clockwise** from the positive x-axis. The terminal side of the angle lies in quadrant IV, so −45° is a quadrant IV angle.

Figure 1.6

Example 3. Sketch a 540° angle in standard position and classify it by quadrant or as a quadrantal angle.

Solution 3.

Since 360° is one full revolution, the positive angle 540° is formed by

$$\frac{540}{360} = 1\frac{1}{2}$$

revolutions **counterclockwise** from the positive x-axis.

The terminal side of the angle lies along the x-axis, so 540° is a quadrantal angle.

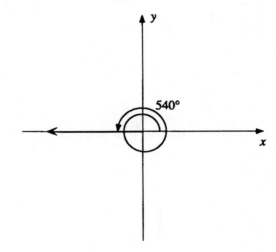

Figure 1.7

Example 4. Estimate the degree measure of the angle θ shown in the sketch and classify the angle by quadrant or as a quadrantal angle.

Solution 4.

The angle θ was formed by rotating the ray **counterclockwise** through $1\frac{1}{3}$ revolutions from the positive x-axis. Therefore, the degree measure of the angle is

$$\left(1\frac{1}{3}\right) \cdot 360° = 480°.$$

The terminal side of the angle lies in quadrant II, so θ is a quadrant II angle.

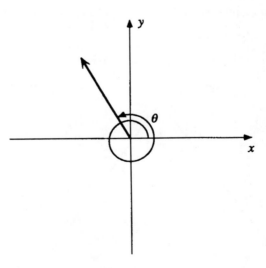

Figure 1.8

12

Example 5. Estimate the degree measure of the angle θ shown in the sketch and classify the angle by quadrant or as a quadrantal angle.

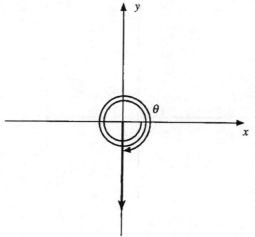

Solution 5.

The angle θ was formed by rotating the ray **clock-wise** through $2\frac{1}{4}$ revolutions from the positive x-axis. Therefore, the degree measure of the angle is

$$\left(-2\frac{1}{4}\right) \cdot 360° = -810°.$$

The terminal side of the angle lies along the y-axis so θ is a quadrantal angle.

Figure 1.9

Practice Problems

In **Problems 1 - 4**, label the **vertex**, the **initial side** and the **terminal side** of the angle shown.

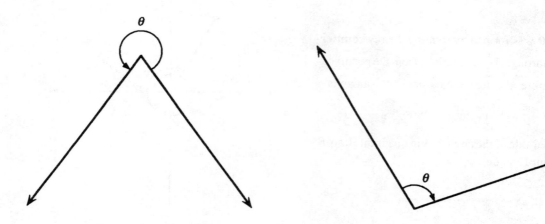

1. **Figure 1.10**

2. **Figure 1.11**

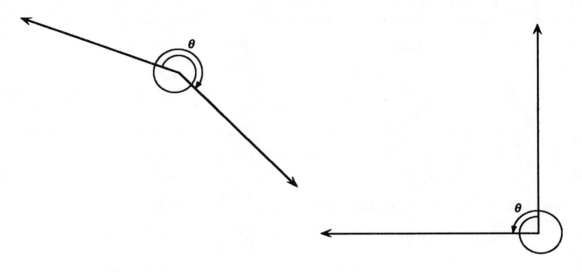

3. Figure 1.12

4. Figure 1.13

In **Problems 5 - 8**, draw and label a rectangular coordinate system on each angle so the angle is in **standard position** in that coordinate system.

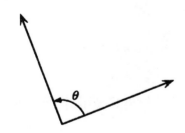

5. Figure 1.14

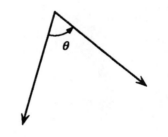

6. Figure 1.15

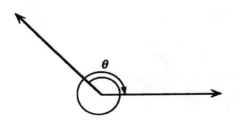

7. Figure 1.16

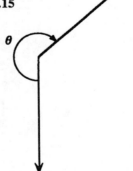

8. Figure 1.17

In **Problems 9 - 14**, classify each angle as an acute angle, an obtuse angle, or a right angle. Remember, this classification is used only for angles between 0° and 180°. These are the angles that occur in triangles.

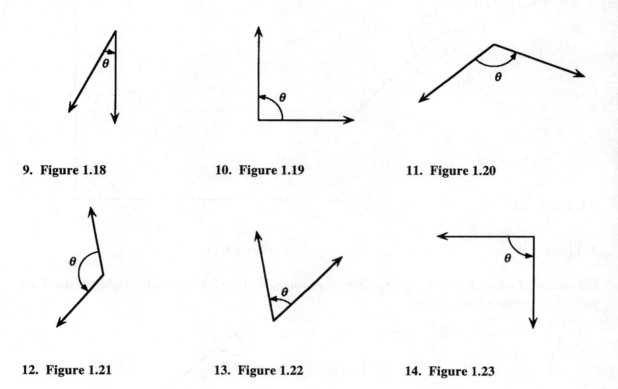

9. Figure 1.18   **10. Figure 1.19**   **11. Figure 1.20**

12. Figure 1.21   **13. Figure 1.22**   **14. Figure 1.23**

In **Problems 15 - 34**, sketch the given angle in standard position. Classify each angle by quadrant or as a quadrantal angle.

15. 60°	**22.** −270°	**29.** −90°
16. 150°	**23.** 637°	**30.** −370°
17. 225°	**24.** 493°	**31.** −315°
18. 330°	**25.** −427°	**32.** 95°
19. 180°	**26.** 575°	**33.** −190°
20. −15°	**27.** −1440°	**34.** −540°
21. −120°	**28.** 255°	

In **Problems 35 - 44**, estimate the degree measure of each angle. Classify each angle by quadrant or as a quadrantal angle.

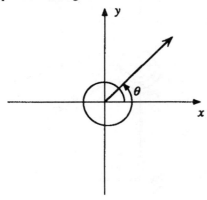

35. Figure 1.24

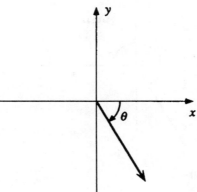

36. Figure 1.25

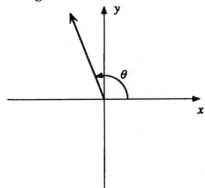

37. Figure 1.26

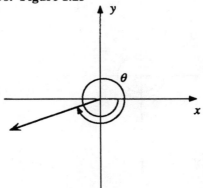

38. Figure 1.27

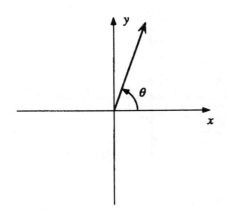

39. Figure 1.28

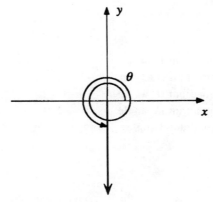

40. Figure 1.29

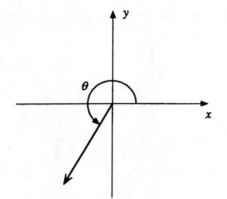

41. Figure 1.30

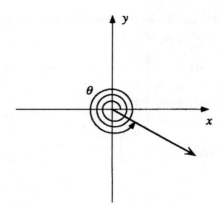

42. Figure 1.31

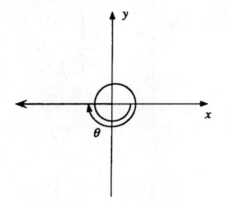

43. Figure 1.32

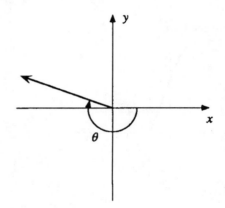

44. Figure 1.33

45. The angle in Figure 1.34 is classified as a

(A) quadrant I angle
(B) quadrant II angle
(C) quadrant III angle
(D) quadrant IV angle
(E) quadrantal angle

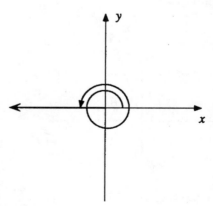

Figure 1.34

46. A 310° angle in standard position is classified as a

   (A) quadrant I angle
   (B) quadrant II angle
   (C) quadrant III angle
   (D) quadrant IV angle
   (E) a quadrantal angle

47. Which one of Figures 1.35 - 1.39 below, shows an angle θ of 75° in standard position?

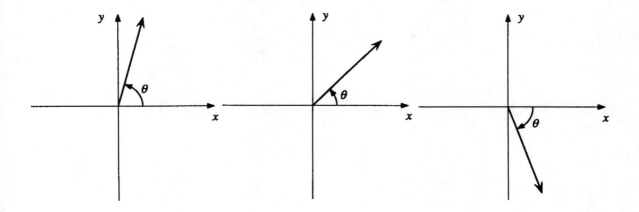

(A) Figure 1.35   **(B) Figure 1.36**   **(C) Figure 1.37**

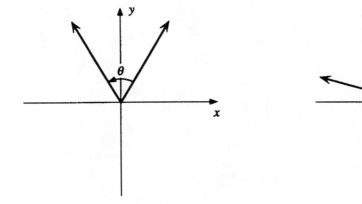

(D) Figure 1.38   **(E) Figure 1.39**

48. What is the approximate degree measure of the angle shown in Figure 1.40?

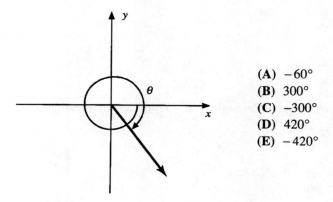

(A) $-60°$
(B) $300°$
(C) $-300°$
(D) $420°$
(E) $-420°$

Figure 1.40

Objective 1.2

Given the degree measure of an angle, find its radian measure. Conversely, given the radian measure of an angle, find its degree measure.

Discussion

The radian is another unit used to describe the size of angles. In analytic trigonometry and calculus this unit is almost always used because it simplifies many formulas.

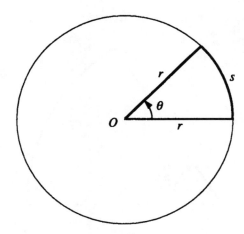

Figure 1.41 shows an angle θ. A circle of radius r has been drawn using the vertex O of the angle θ as its center. On the circumference of the circle the angle θ cuts off an arc labeled s. The **radian measure** of the angle θ is defined as the length of the arc s that the angle cuts off on the circumference of the circle measured in *radius units*.

Figure 1.41

Because it is convenient, we customarily use the same symbol to denote a line segment and its length, or an arc and its length, or an angle and its measure. Thus, in Figure 1.41, r might denote either a radius of the circle (a line segment) or the length of a radius; s might denote either the arc of the circle or the length of the arc; and θ might denote either the angle or its measure (in whatever unit of measure is being used). With this understanding about notation, θ, r and s are related by the equations

(1)
$$\theta \text{ (in radians)} = \frac{s}{r},$$

or

(2)
$$s = r\theta \ (\theta \text{ in radians}).$$

For the angle formed by one full revolution, the arc length s is equal to the circumference of the circle. From the familiar formula for the circumference of a circle, in this situation

$$s = 2\pi r$$

and

$$\frac{s}{r} = 2\pi .$$

Thus, from equation (1), an angle of one full revolution has radian measure 2π.

Since one full revolution is $360°$, we obtain the basic relationship

(3)
$$\boxed{360° = 2\pi \text{ radians.}}$$

On dividing equation (3) by 2π, we discover that 1 radian $= \dfrac{360°}{2\pi} \approx 57.3°$ (where the symbol $\approx$ means *is approximately equal to*).

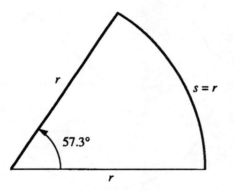

From equation (1) we also see that an angle of **one radian** is formed when the arc length s equals the radius r. Thus, an angle of $57.3°$ (slightly less than $\dfrac{1}{6}$ revolution) cuts off an arc of the circle of length approximately equal to the radius of the circle.

Figure 1.42 An angle of one radian

Since $360° = 2\pi$ radians, we understand that

$$180° = \frac{1}{2} \cdot 360° = \frac{1}{2} \cdot 2\pi = \pi \text{ radians}, \qquad 60° = \frac{1}{6} \cdot 360° = \frac{\pi}{3} \text{ radians},$$

$$120° = \frac{1}{3} \cdot 360° = \frac{2\pi}{3} \text{ radians}, \qquad 45° = \frac{1}{8} \cdot 360° = \frac{\pi}{4} \text{ radians},$$

$$90° = \frac{1}{4} \cdot 360° = \frac{\pi}{2} \text{ radians} \qquad \text{and} \qquad 30° = \frac{1}{12} \cdot 360° = \frac{\pi}{6} \text{ radians}.$$

These special angles occur frequently and their degree-radian conversions should be memorized.

In general, to convert between degrees and radians, we use the equations

(4)
$$1° = \frac{\pi}{180} \text{ radians and } 1 \text{ radian} = \frac{180°}{\pi}.$$

Summary

1. The radian measure of an angle θ is the length of the arc s that the angle θ subtends on a circle of any radius r centered at the vertex of the angle, measured in radius units. Thus,

$$\theta \text{ (in radians)} = \frac{s}{r} \text{ and } s = r\theta.$$

2. An angle of one revolution counterclockwise $= 2\pi$ radians $= 360°$.
Therefore,

$$1 \text{ radian} = \frac{360°}{2\pi} \text{ is approximately } 57.3°.$$

3. An angle of

- one-half revolution counterclockwise $= 180° = \frac{2\pi}{2}$ radians $= \pi$ radians,

- one-third revolution counterclockwise $= 120° = \frac{2\pi}{3}$ radians,

- one-fourth revolution counterclockwise $= 90° = \frac{2\pi}{4}$ radians $= \frac{\pi}{2}$ radians,

- one-sixth revolution counterclockwise $= 60° = \frac{2\pi}{6}$ radians $= \frac{\pi}{3}$ radians,

- one-eighth revolution counterclockwise $= 45° = \frac{2\pi}{8}$ radians $= \frac{\pi}{4}$ radians,

- one-twelfth revolution counterclockwise $= 30° = \frac{2\pi}{12}$ radians $= \frac{\pi}{6}$ radians.

These angles occur frequently in trigonometry.

4. To convert between degrees and radians, use the equations

$$1° = \frac{\pi}{180} \text{ radians and } 1 \text{ radian} = \frac{180°}{\pi}.$$

Examples The following Examples illustrate the conversion process.

Example 1. Convert 330° into radians.

Solution 1.

Since $1° = \dfrac{\pi}{180}$ radians, we have

$$330° = 330\left(\frac{\pi}{180}\right) \text{ radians } = \frac{11\pi}{6} \text{ radians.}$$

When expressing the measure of an angle in radians, the word *radian* is customarily omitted, so this answer would be written simply as

$$330° = \frac{11\pi}{6}.$$

Example 2. Convert −720° into radians.

Solution 2.

Since $1° = \dfrac{\pi}{180}$ radians, we have

$$-720° = -720\left(\frac{\pi}{180}\right) = -4\pi.$$

Example 3. Convert $\dfrac{3\pi}{4}$ radians into degrees.

Solution 3.

Since $1 \text{ radian} = \dfrac{180°}{\pi}$, we have

$$\frac{3\pi}{4} \text{ radians } = \frac{3\pi}{4}\left(\frac{180}{\pi}\right) = 135°.$$

Example 4. Convert 3 radians into degrees. Express your answer to the nearest tenth of a degree.

Solution 4.

Since $1 \text{ radian} = \dfrac{180°}{\pi}$, we have

$$3 \text{ radians } = 3\left(\frac{180}{\pi}\right) = 171.9°.$$

This calculation was done on a scientific calculator. Since 171.9° is an approximation for 3 radians, it would be more accurate to use the symbol ≈ instead of = . For simplicity, however, we will use the symbol = in our calculations.

Practice Problems

In **Problems 1 - 9**, convert the given degree measures into exact radian measures.

1. 150°   **4.** 540°   **7.** −450°

2. 300°   **5.** 75°   **8.** 217°

3. −225°   **6.** −60°   **9.** −765°

In **Problems 10 - 15**, convert the given degree measures into radian measures. Express your answers to the nearest hundredth of a radian.

10. 23°   **13.** 247°

11. −215°   **14.** 900°

12. 172°   **15.** −81°

In **Problems 16 - 27**, convert the given radian measures into degree measures. Express your answers to the nearest tenth of a degree.

16. $\dfrac{\pi}{3}$ **19.** $\dfrac{7\pi}{6}$ **22.** −5.50   **25.** 13

17. 3π **20.** 4π **23.** −1.05   **26.** −2.7

18. $-\dfrac{5\pi}{4}$ **21.** 1   **24.** 0.5   **27.** 1.57

28. What is the radian measure of a 192° angle expressed to the nearest hundredth of a radian?

(A) 3.33   **(B)** 3.35   **(C)** 3.37   **(D)** 3.39   **(E)** 3.41

29. The radian measure of an angle is $\dfrac{4}{7}\pi$. What is its degree measure, expressed to the nearest tenth of a degree?

(A) 102.3°   **(B)** 102.9°   **(C)** 103.5°   **(D)** 104.1°   **(E)** 104.7°

Objective 1.3

Given a point on the terminal side of an angle in standard position, evaluate the trigonometric functions for the angle.

Discussion

In numerical trigonometry, the six trigonometric functions are used to describe the relationships among the sides and angles of triangles. To define these functions, consider an angle θ in standard position as in Figure 1.43.

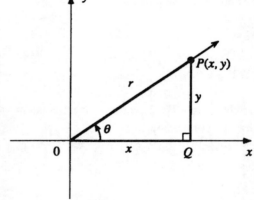

Choose any point P (other than the origin) with coordinates (x, y) on the terminal side of θ. From P construct a perpendicular to the x-axis at Q, thus forming right triangle OQP. The length of the hypotenuse of this right triangle is labeled r. By the Pythagorean Theorem

$$r = \sqrt{x^2 + y^2} \text{ (positive square root).}$$

Figure 1.43

The coordinates x and y of point P can be positive, negative, or zero, depending on where the terminal side of angle θ lies. For example, if θ is a quadrant III angle, then $x < 0$ and $y < 0$. However, we always have $r > 0$.

In terms of the quantities x, y and r, the six **trigonometric functions** of the angle θ are defined by

(5)

$$
\begin{aligned}
&\text{sine } \theta = \sin \theta = \frac{y}{r}, && \text{cosecant } \theta = \csc \theta = \frac{r}{y}, \\
&\text{cosine } \theta = \cos \theta = \frac{x}{r}, && \text{secant } \theta = \sec \theta = \frac{r}{x}, \\
&\text{tangent } \theta = \tan \theta = \frac{y}{x}, && \text{cotangent } \theta = \cot \theta = \frac{x}{y}.
\end{aligned}
$$

Note that these definitions do not depend on the point $P(x, y)$ used above. Indeed, suppose we choose another point $P'(x', y')$ on the terminal side of θ as in Figure 1.44.

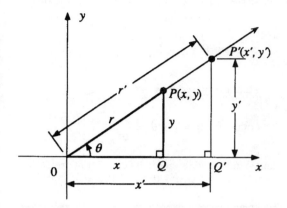

Set $r' = \sqrt{(x')^2 + (y')^2}$. Then the two right triangles OQP and $OQ'P'$ are similar.
Hence, by similar triangles,

$$\frac{y}{r} = \frac{y'}{r'} = \sin \theta.$$

Figure 1.44

It follows that the same value of $\sin \theta$ is obtained no matter what point $P(x, y)$ on the terminal side of θ is used for the calculation. The same is true for the other five trigonometric functions.

Because $r = \sqrt{x^2 + y^2}$, $|x| \leq r$ and $|y| \leq r$. Consequently, for any angle θ , $|\sin \theta| \leq 1$, $|\cos \theta| \leq 1$, $|\csc \theta| \geq 1$ and $|\sec \theta| \geq 1$. Thus, the values of the sine and cosine lie between -1 and 1, and the values of the cosecant and the secant are larger than 1 and smaller than -1. The coordinates x and y can be of any size and are independent of each other, so the tangent and the cotangent can have any values.

Since $\dfrac{1}{\sin \theta} = \dfrac{1}{y \big/ r} = \dfrac{r}{y} = \csc \theta$, sine and cosecant are called **reciprocal functions**. Similarly, cosine and secant are reciprocal functions and tangent and cotangent are reciprocal functions.

Authors' Note

Trigonometry originated in the work of the Egyptians and Babylonians about 4,000 years ago. The ancient Greeks developed and/or used theorems from which began the evolution of modern day trigonometric laws. The medieval European, Hindu and Islamic cultures used their acquired knowledge of Hellenic trigonometry to develop a new form of the study. The use of all six of the common trigonometric functions, and the relations among them, appeared at this time. Over the centuries trigonometry has grown and evolved into its present-day form, a form in which it is primarily the study and application of the six trigonometric functions: the sine, cosecant, cosine, secant, tangent and cotangent.

26

Summary

1. The six trigonometric functions of an angle θ in standard position are defined in terms of any point (x, y) on the terminal side of the angle and the distance $r = \sqrt{x^2 + y^2}$ of this point from the origin by these formulas:

$$\text{sine } \theta = \sin \theta = \frac{y}{r}, \qquad\qquad \text{cosecant } \theta = \csc \theta = \frac{r}{y},$$

$$\text{cosine } \theta = \cos \theta = \frac{x}{r}, \qquad\qquad \text{secant } \theta = \sec \theta = \frac{r}{x},$$

$$\text{tangent } \theta = \tan \theta = \frac{y}{x}, \qquad\qquad \text{cotangent } \theta = \cot \theta = \frac{x}{y}.$$

2. The pairs of functions sine and cosecant, cosine and secant, and tangent and cotangent are called **reciprocal functions**.

Examples

Example 1. An angle θ in standard position has the point $(-3, 5)$ on its terminal side. Evaluate the trigonometric functions for θ.

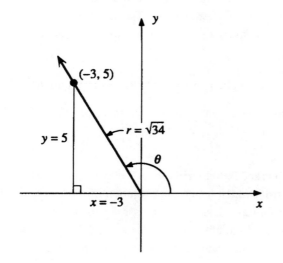

Solution 1.

First, we sketch θ in standard position using the fact that $(-3, 5)$ lies on its terminal side. See Figure 1.45. Notice that θ is a quadrant II angle.

Second, choose the point $P(-3, 5)$ on the terminal side of θ so $x = -3$ and $y = 5$. Use $r = \sqrt{x^2 + y^2}$ to compute $r = \sqrt{(-3)^2 + (5)^2} = \sqrt{34}$.

Figure 1.45

Third, use the values $x = -3$, $y = 5$, $r = \sqrt{34}$ to evaluate the trigonometric functions for θ from their definitions given by equations (5).

$$\sin \theta = \frac{y}{r} = \frac{5}{\sqrt{34}} = \frac{5\sqrt{34}}{34} \qquad\qquad \csc \theta = \frac{r}{y} = \frac{\sqrt{34}}{5}$$

$$\cos \theta = \frac{x}{r} = \frac{-3}{\sqrt{34}} = \frac{-3\sqrt{34}}{34} \qquad\qquad \sec \theta = \frac{r}{x} = \frac{\sqrt{34}}{-3}$$

$$\tan \theta = \frac{y}{x} = \frac{5}{-3} \qquad\qquad \cot \theta = \frac{x}{y} = \frac{-3}{5}$$

Example 2. An angle θ in standard position has the point $(0, -2)$ on its terminal side. Evaluate the trigonometric functions for θ .

Solution 2.
In sketching θ we find that it is a quadrantal angle with the point $(0, -2)$ on its terminal side. See Figure 1.46. Choose the point $P(0, -2)$ on the terminal side of θ so $x = 0$, $y = -2$ and

$$r = \sqrt{x^2 + y^2} = \sqrt{0^2 + (-2)^2} = 2 .$$

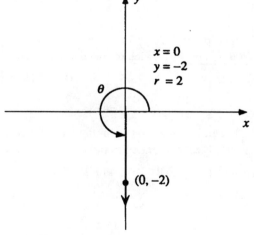

$$x = 0$$
$$y = -2$$
$$r = 2$$

$$(0, -2)$$

Figure 1.46

Finally, use $x = 0$, $y = -2$, $r = 2$ to evaluate the six trigonometric functions from their definitions.

$$\sin \theta = \frac{-2}{2} = -1 \qquad\qquad \csc \theta = \frac{2}{-2} = -1$$

$$\cos \theta = \frac{0}{2} = 0 \qquad\qquad \sec \theta = \frac{2}{0} \text{ (undefined)}$$

$$\tan \theta = \frac{-2}{0} \text{ (undefined)} \qquad\qquad \cot \theta = \frac{0}{-2} = 0$$

The tangent and secant of this angle are undefined because division by zero is not defined.

Practice Problems

In **Problems 1 - 15**, a point is given on the terminal side of an angle θ in standard position. Evaluate the six trigonometric functions of θ .

1. $(3, 4)$

2. $(-5, 12)$

3. $(-24, -7)$

4. $\left(\sqrt{3}, -1\right)$

5. $(-2, 0)$

6. $(7, 1)$

7. $(-10, 7)$

8. $(0, -5)$

9. $\left(\frac{2}{3}, 1\right)$

10. $\left(-\frac{1}{3}, -\frac{1}{4}\right)$

11. $\left(-\frac{1}{3}, \frac{3}{4}\right)$

12. $\left(-3, \sqrt{7}\right)$

13. $\left(\frac{1}{2}, \frac{\sqrt{3}}{2}\right)$

14. $\left(\sqrt{2}, \sqrt{10}\right)$

15. $\left(\frac{2}{3}, -\sqrt{2}\right)$

16. The point $(5, -7)$ is on the terminal side of angle θ in standard position. What is the value of $3 \sin \theta - 8 \cot \theta$ rounded to two decimal places?

(A) 3.27   **(B)** 7.46   **(C)** 8.16   **(D)** 8.76   **(E)** 12.94

17. The point $(-9, -4)$ is on the terminal side of an angle θ in standard position. What is the value of $2 \tan \theta - 5 \sec \theta$ rounded to three decimal places?

(A) −7.811   **(B)** −6.360   **(C)** −4.583   **(D)** 6.360   **(E)** 7.811

Objective 1.4

Given information which determines the quadrant of an angle and given the value of one trigonometric function of that angle, find the values of the six trigonometric functions of that angle.

Discussion

Recall that

$$\sin \theta = \frac{y}{r}, \qquad\qquad \csc \theta = \frac{r}{y},$$

$$\cos \theta = \frac{x}{r}, \qquad\qquad \sec \theta = \frac{r}{x},$$

$$\tan \theta = \frac{y}{x}, \qquad\qquad \cot \theta = \frac{x}{y},$$

where (x, y) is any point on the terminal side of θ and $r = \sqrt{x^2 + y^2}$. Since r is always positive, the signs of x and y determine the signs of the trigonometric functions. Therefore, $\sin \theta = \frac{y}{r}$ is positive when $y > 0$ and negative when $y < 0$. Thus, $\sin \theta$ is positive if and only if the terminal side of θ lies in the upper half of the coordinate plane. Since $\sin \theta$ and $\csc \theta$ always have the same sign, this is also true of the cosecant. Analogously, $\cos \theta = \frac{x}{r}$ is positive when $x > 0$ and negative when $x < 0$. Thus, $\cos \theta$ is positive if and only if the terminal side of θ lies in the right half of the coordinate plane. Since $\cos \theta$ and $\sec \theta$ always have the same sign, this is also true of the secant.

The signs of $\tan \theta$ and $\cot \theta$ are always the same and depend on the signs of both x and y. These functions are positive when x and y have the same sign and negative when x and y have different signs. Thus, $\tan \theta$ and $\cot \theta$ are positive when the terminal side of θ lies in either quadrant I (where x and y are both positive) or quadrant III (where x and y are both negative), and are negative when the terminal side lies in quadrant II or quadrant IV.

Figure 1.47 shows how the signs of x, y and r and of the six trigonometric functions depend on the quadrant.

Quadrant II
$x < 0$, $y > 0$, $r > 0$

$\sin\theta = \dfrac{y}{r} > 0$ and $\csc\theta = \dfrac{r}{y} > 0$

$\cos\theta = \dfrac{x}{r} < 0$ and $\sec\theta = \dfrac{r}{x} < 0$

$\tan\theta = \dfrac{y}{x} < 0$ and $\cot\theta = \dfrac{x}{y} < 0$

sine and cosecant positive

Quadrant I
$x > 0$, $y > 0$, $r > 0$

$\sin\theta = \dfrac{y}{r} > 0$ and $\csc\theta = \dfrac{r}{y} > 0$

$\cos\theta = \dfrac{x}{r} > 0$ and $\sec\theta = \dfrac{r}{x} > 0$

$\tan\theta = \dfrac{y}{x} > 0$ and $\cot\theta = \dfrac{x}{y} > 0$

All positive

Quadrant III
$x < 0$, $y < 0$, $r > 0$

$\sin\theta = \dfrac{y}{r} < 0$ and $\csc\theta = \dfrac{r}{y} < 0$

$\cos\theta = \dfrac{x}{r} < 0$ and $\sec\theta = \dfrac{r}{x} < 0$

$\tan\theta = \dfrac{y}{x} > 0$ and $\cot\theta = \dfrac{x}{y} > 0$

tangent and cotangent positive

Quadrant IV
$x > 0$, $y < 0$, $r > 0$

$\sin\theta = \dfrac{y}{r} < 0$ and $\csc\theta = \dfrac{r}{y} < 0$

$\cos\theta = \dfrac{x}{r} > 0$ and $\sec\theta = \dfrac{r}{x} > 0$

$\tan\theta = \dfrac{y}{x} < 0$ and $\cot\theta = \dfrac{x}{y} < 0$

cosine and secant positive

Figure 1.47

The quadrant of an angle can be determined from the signs of two trigonometric functions which are not a reciprocal pair. For instance, if $\sin\theta > 0$ and $\cot\theta < 0$, then θ must be a quadrant II angle because only in quadrant II is the sine positive and the cotangent negative.

The values of all of the trigonometric functions can be found from the quadrant of an angle θ and the value of one of the trigonometric functions for θ. If the quadrant of θ is known, a point on the terminal side of θ can be found from the value of any one of the trigonometric functions. From the coordinates of this point, the values of all the trigonometric functions can be calculated.

To see why the value of one trigonometric function does not, by itself, determine the value of the others, suppose that $\tan \theta = \dfrac{p}{q}$. This value could arise, for instance, from either the point (p, q) or the point $(-p, -q)$ being on the terminal side of θ. Additional information (such as the sign of another trigonometric function other than the cotangent) determines the quadrant of θ and allows us to choose between these points. If (p, q) is found to be a point on the terminal side of θ, compute $r = \sqrt{p^2 + q^2}$ and use p, q and r to compute the values of the remaining trigono-metric functions.

As another illustration, suppose that $\sin \theta = \dfrac{p}{q}$. This value could arise, for instance, from any point (x, y) for which $x = p$ and $r = q = \sqrt{x^2 + y^2}$. Thus, this value for $\sin \theta$ could arise from either the point $\left(p, +\sqrt{q^2 - p^2}\right)$ or the point $\left(p, -\sqrt{q^2 - p^2}\right)$. Additional information, such as the sign of another trigonometric function (other than the cosecant), determines the quadrant of θ and allows us to choose between these points. If $\left(p, -\sqrt{q^2 - p^2}\right)$ is found to be a point on the terminal side of θ, we can use $x = p$, $y = -\sqrt{q^2 - p^2}$ and $r = q$ to compute the values of the remaining trigonometric functions.

Summary

1. The quadrant of an angle θ can be determined from the signs of two trigonometric functions which are not a pair of reciprocal functions.

2. Once the quadrant of an angle θ is known, the coordinates of a point on the terminal side of θ can be computed from the value of any one of the trigonometric functions. The values of all the trigonometric functions can be computed from these coordinates.

Examples

Example 1. Identify the possible quadrant(s) of angle θ if $\tan \theta < 0$.

Solution 1.
We see from Figure 1.47 that $\tan \theta$ is positive in quadrants I and III and negative in quadrants II and IV. Thus, θ is either a quadrant II or a quadrant IV angle.

Example 2. Identify the quadrant(s) of angle ϕ if $\cos \phi < 0$ and $\cot \phi > 0$.

Solution 2.
The cosine is negative for quadrant II and quadrant III angles. The cotangent is positive for quadrant I and quadrant III angles. Only in quadrant III is $\cos \phi < 0$ and $\cot \phi > 0$. Thus, ϕ is a quadrant III angle.

Example 3. Angle θ is a quadrant III angle and $\sin \theta = \dfrac{-3}{7}$. Find the values of all six trigonometric functions of θ.

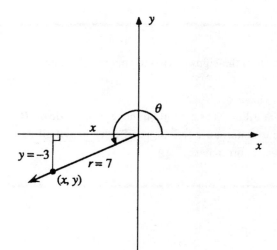

Figure 1.48

Solution 3.
First, sketch the angle θ in quadrant III as in Figure 1.48. Second, choose a point (x, y) on the terminal side of θ so that

$$\sin \theta = \frac{y}{r} = \frac{-3}{7}.$$

Remember that r is always positive, and that y is negative in quadrant III. It is easiest to choose this point so $y = -3$ and $r = 7$. Mark this point in the figure.

Third, compute the x-coordinate of this point by using $x^2 + y^2 = r^2$.

$$x^2 + (-3)^2 = (7)^2$$
$$x^2 = 40$$
$$x = \sqrt{40} \text{ or } x = -\sqrt{40}.$$

Since the point is in quadrant III, x must be negative so $x = -\sqrt{40}$.
Finally, evaluate the six trigonometric functions from their definitions using $x = -\sqrt{40}$, $y = -3$ and $r = 7$. We obtain

$$\sin \theta = \frac{-3}{7}, \qquad\qquad \csc \theta = \frac{7}{-3} = -\frac{7}{3},$$

$$\cos \theta = \frac{-\sqrt{40}}{7}, \qquad\qquad \sec \theta = \frac{7}{-\sqrt{40}} = -\frac{7}{\sqrt{40}},$$

$$\tan \theta = \frac{-3}{-\sqrt{40}} = \frac{3\sqrt{40}}{40}, \qquad\qquad \cot \theta = \frac{-\sqrt{40}}{-3} = \frac{\sqrt{40}}{3}.$$

Example 4. Angle θ is in standard position, $\cos \theta < 0$ and $\tan \theta = \frac{-1}{2}$. Find the values of all the trigonometric functions at θ.

Solution 4.
First, determine the quadrant of θ and sketch the angle. Since $\cos \theta < 0$ and $\tan \theta < 0$, we see from Figure 1.47 that θ is a quadrant II angle. Since θ is a quadrant II angle and

$$\tan \theta = \frac{-1}{2},$$

we can sketch θ as in Figure 1.49.
Second, choose a point (x, y) on the terminal side of θ so that $x < 0, \quad y > 0$ and

$$\tan \theta = \frac{y}{x} = -\frac{1}{2}.$$

The easiest choice is $(-2, 1)$.

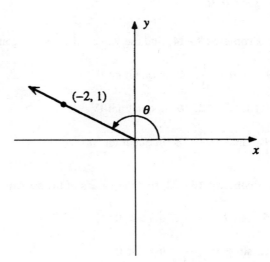

Figure 1.49

Third, compute $r = \sqrt{x^2 + y^2} = \sqrt{(-2)^2 + (1)^2} = \sqrt{5}$.

Finally, evaluate the trigonometric functions from their definitions using $x = -2$, $y = 1$, and $r = \sqrt{5}$. We obtain

$$\sin \theta = \frac{y}{r} = \frac{1}{\sqrt{5}} = \frac{\sqrt{5}}{5}, \qquad \csc \theta = \frac{r}{y} = \frac{\sqrt{5}}{1} = \sqrt{5},$$

$$\cos \theta = \frac{x}{r} = \frac{-2}{\sqrt{5}} = \frac{-2\sqrt{5}}{5}, \qquad \sec \theta = \frac{r}{x} = -\frac{\sqrt{5}}{2},$$

$$\tan \theta = \frac{y}{x} = \frac{-1}{2}, \qquad \cot \theta = \frac{x}{y} = -2.$$

Practice Problems

In **Problems 1 - 8**, identify the quadrant(s) of an angle θ in standard position satisfying the given condition(s).

1. $\cos \theta > 0$

2. $\sin \theta < 0$

3. $\csc \theta > 0$

4. $\sec \theta < 0$

5. $\sin \theta > 0$ and $\cos \theta < 0$

6. $\tan \theta < 0$ and $\sec \theta > 0$

7. $\csc \theta > 0$ and $\cot \theta < 0$

8. $\sin \theta < 0$ and $\cot \theta > 0$

In **Problems 9 - 14**, find the values of the six trigonometric functions of the angle.

9. $\sin \theta = \frac{1}{2}$, θ in quadrant II

10. $\sec \theta = 2$, θ in quadrant IV

11. $\tan \phi = \frac{1}{3}$, ϕ in quadrant III

12. $\cot \alpha = 6$, α in quadrant I

13. $\cos \beta = -\frac{5}{13}$, β in quadrant III

14. $\csc \gamma = -\frac{7}{2}$, γ in quadrant IV

In **Problems 15 - 22**, find the values of the six trigonometric functions of the angle.

15. $\cos \theta = -\frac{8}{17}$, $\sin \theta > 0$

16. $\sin \phi = -\frac{12}{13}$, $\tan \phi < 0$

17. $\cot \theta = \frac{15}{8}$, $\sec \theta < 0$

18. $\sec \phi = -\frac{5}{2}$, $\cot \phi < 0$

19. $\sin \beta = -\frac{5}{12}$, $\cot \beta > 0$

20. $\tan \alpha = 2$, $\sin \alpha > 0$

21. $\cos \theta = \frac{-2}{7}$, $\sin \theta < 0$

22. $\csc \phi = \frac{5}{2}$, $\tan \phi > 0$

23. Angle θ is a quadrant II angle and $\sin \theta = \dfrac{1}{3}$. What is the value of $\underline{3 \tan \theta - \sec \theta}$, rounded to two decimal places?

(A) -7.42 (B) -4.06 (C) -1.94 (D) 2.00 (E) 0.00

24. Tan $\theta = -4$ and $\sec \theta$ is positive. What is the value of $\underline{7 \cos \theta + 3 \csc \theta}$, rounded to one decimal place?

(A) -1.2 (B) -1.4 (C) 4.8 (D) 5.6 (E) 14.1

UNIT 1
Sample Examination 1

1. What is the approximate degree measure of the angle shown?

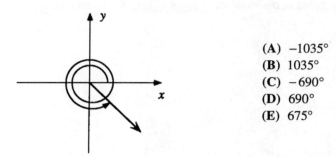

(A) −1035°
(B) 1035°
(C) −690°
(D) 690°
(E) 675°

Figure 1.50

2. What is the degree measure of an angle of −0.55 radians, expressed to the nearest hundredth of a degree?

(A) −63.06°   (B) −63.03°   (C) −62.70°   (D) −31.53°   (E) −31.51°

3. The point $(9, 6)$ is on the terminal side of angle θ in standard position. What is the value of $8 \tan \theta + 5 \csc \theta$, rounded to one decimal place?

(A) 8.1   (B) 8.6   (C) 11.3   (D) 14.3   (E) 17.8

4. Angle θ is a quadrant III angle and $\tan \theta = \dfrac{8}{5}$. What is the value of $7 \sin \theta + 8 \sec \theta$, rounded to three decimal places?

(A) −21.030   (B) −18.804   (C) −15.370   (D) −13.144   (E) −10.176

5. $\cot \theta = \dfrac{11}{2}$ and $\sin \theta$ is negative. What is the value of $5 \cos \theta + 7 \csc \theta$, rounded to three decimal places?

(A) −8.009   (B) −12.034   (C) −34.212   (D) −44.051   (E) −44.213

UNIT 1
Sample Examination 2

1. Which figure shows a −525° angle in standard position?

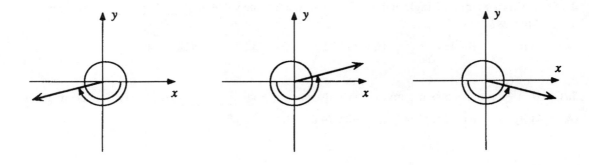

(A) **Figure 1.51** (B) **Figure 1.52** (C) **Figure 1.53**

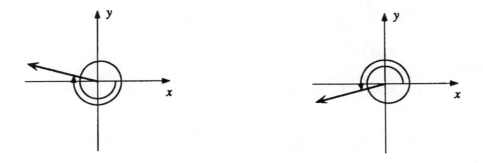

(D) **Figure 1.54** (E) **Figure 1.55**

2. What is the radian measure of a $-792°$ angle expressed to the nearest tenth of a radian?

(A) -13.2 (B) -13.8 (C) -13.9 (D) -14.1 (E) -14.3

3. The point $(6, -5)$ is on the terminal side of angle θ in standard position. What is the value of $\sin \theta - 3 \sec \theta$, rounded to one decimal place?

(A) -4.5 (B) -3.1 (C) -2.9 (D) 3.3 (E) 5.5

4. Angle θ is a quadrant I angle and $\sin \theta = \dfrac{3}{14}$. What is the value of $3 \cos \theta + 5 \cot \theta$, rounded to three decimal places?

(A) 28.116 (B) 25.962 (C) 25.722 (D) 4.027 (E) 1.740

5. $\cos \theta = \dfrac{2}{7}$ and $\sin \theta$ is negative. What is the value of $8 \tan \theta + 2 \csc \theta$, rounded to four decimal places?

(A) -28.9198 (B) -28.7494 (C) -24.7458 (D) -19.8328 (E) -4.4721

UNIT 2

EVALUATING TRIGONOMETRIC FUNCTIONS

Introduction

In this Unit we will study the numerical evaluation of the trigonometric functions. First, exact values will be obtained for the quadrantal angles 0°, 90°, 180°, 270°, and 360°. Second, using scientific calculators, approximate values will be calculated for arbitrary angles. The reference angle is introduced to solve the inverse problem: given a trigonometric function value, find the angle(s). Both degree measure and radian measure will be used throughout.

UNIT 2
EVALUATING TRIGONOMETRIC FUNCTIONS

Objective 2.1

State the values of the trigonometric functions for the quadrantal angles $0°$, $90°$, $180°$, $270°$, and $360°$ $\left(0, \dfrac{\pi}{2}, \pi, \dfrac{3\pi}{2} \text{ and } 2\pi \text{ radians}\right)$.

Objective 2.2

Given the degree or radian measure of an angle, use a scientific calculator to find approximate values of the trigonometric functions for the angle.

Objective 2.3

Given an angle between $0°$ and $360°$ (0 and 2π radians), find its reference angle.

Objective 2.4

Given an angle between $0°$ and $360°$ (0 and 2π radians), express the values of the trigonometric functions for the given angle in terms of its reference angle.

Objective 2.5

(a) Given a numerical value $a \geq 0$ for $\sin \theta$, $\cos \theta$, or $\tan \theta$, find all angles between $0°$ and $90°$ (0 and $\dfrac{\pi}{2}$ radians) which have the given function value or state that there is no such angle.

(b) Given a numerical value for $\sin \theta$, $\cos \theta$, or $\tan \theta$, find all angles between $0°$ and $360°$ (0 and 2π radians) which have the given function value or state that there is no such angle (*i.e.*, solve the inverse problem for the sine, cosine and tangent).

EVALUATING TRIGONOMETRIC FUNCTIONS

Objective 2.1

State the values of the trigonometric functions for the quadrantal angles $0°$, $90°$, $180°$, $270°$, **and** $360°$ $(0,$ $\dfrac{\pi}{2}$, π, $\dfrac{3\pi}{2}$ **and** 2π **radians).**

Discussion

Figures 2.1 (a) - 2.1 (d) show the quadrantal angles in standard position.

$$0°(0 \text{ radians}) \qquad 180°(\pi \text{ radians})$$

$$90°\left(\frac{\pi}{2} \text{ radians}\right) \qquad 270° \left(\frac{3\pi}{2} \text{ radians}\right)$$

$$360°(2\pi \text{ radians})$$

On the terminal side of each angle we have chosen a convenient point.

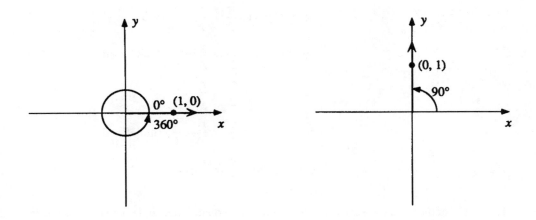

Figure 2.1 (a)
$0°$ and $360°$ quadrantal angles;
$(1, 0)$ on terminal side

Figure 2.1 (b)
$90°$ quadrantal angle;
$(0, 1)$ on terminal side

42

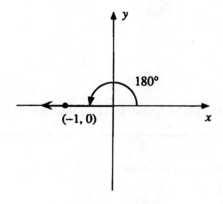

Figure 2.1 (c)
180° quadrantal angle;
$(-1, 0)$ on terminal side

Figure 2.1 (d)
270° quadrantal angle;
$(0, -1)$ on terminal side

For the $90°\left(\dfrac{\pi}{2}\ \text{radians}\right)$ right angle the point chosen has coordinates $x = 0$ and $y = 1$, so

$$r = \sqrt{x^2 + y^2} = \sqrt{0^2 + 1^2} = 1.$$

Therefore, from the definitions of the trigonometric functions,

$$\sin 90° = \sin\frac{\pi}{2} = \frac{y}{r} = \frac{1}{1} = 1,$$

$$\cos 90° = \cos\frac{\pi}{2} = \frac{x}{r} = \frac{0}{1} = 0,$$

$$\tan 90° = \tan\frac{\pi}{2} = \frac{y}{x} = \frac{1}{0}\ \text{(undefined)},$$

$$\csc 90° = \csc\frac{\pi}{2} = \frac{r}{y} = \frac{1}{1} = 1,$$

$$\sec 90° = \sec\frac{\pi}{2} = \frac{r}{x} = \frac{1}{0}\ \text{(undefined)}$$

and

$$\cot 90° = \cot\frac{\pi}{2} = \frac{x}{y} = \frac{0}{1} = 0.$$

The values of the trigonometric functions for the other quadrantal angles can be obtained in the same way. Table 2.1 summaries these results.

In what follows these function values will often be used without comment or explanation. The reader should memorize these function values or learn to use the definitions as above to find these values quickly.

angle θ degrees	angle θ radians	$\sin \theta$	$\cos \theta$	$\tan \theta$	$\csc \theta$	$\sec \theta$	$\cot \theta$
0°	0	0	1	0	undefined	1	undefined
90°	$\dfrac{\pi}{2}$	1	0	undefined	1	undefined	0
180°	π	0	−1	0	undefined	−1	undefined
270°	$\dfrac{3\pi}{2}$	−1	0	undefined	−1	undefined	0
360°	2π	0	1	0	undefined	1	undefined

Table 2.1

In Unit 5 we will apply the definitions of the trigonometric functions in a similar way to obtain the exact values of the trigonometric functions for the special angles 45°, 30° and 60°.

Summary

To find the values of the trigonometric functions for a quadrantal angle, sketch the angle in standard position. Its terminal side will lie along a coordinate axis. Choose the point from among $(1, 0)$, $(-1, 0)$, $(0, 1)$ and $(0, -1)$ on the terminal side of the angle. This point is one unit from the origin, so $r = 1$. Now use the coordinates of the point you chose on the terminal side and $r = 1$ in the definitions to find the values of the trigonometric functions.

Examples

Example 1. Find $\sin 90° + 5 \cos 0°$.

Solution 1.
On substituting the values for $\sin 90°$ and $\cos 0°$, we obtain

$$\sin 90° + 5 \cos 0° = 1 + 5 \cdot 1 = 6.$$

Example 2. Find $\cot \dfrac{3\pi}{2} + \tan \dfrac{\pi}{2}$.

Solution 2.

We know that $\tan \dfrac{\pi}{2}$ is undefined. Thus, $\cot \dfrac{3\pi}{2} + \tan \dfrac{\pi}{2}$ is undefined.

Practice Problems

Evaluate the following expressions without using a calculator.

1. $\sin 90° \cos 180° - \sin 270°$

2. $\tan \dfrac{\pi}{2} - \cot \pi + 3 \sec \dfrac{3\pi}{2}$

3. $\sec 180° \cot 90° \sec 360°$

4. $\tan \pi - \sec \pi + 5 \csc \dfrac{\pi}{2}$

5. $3 \tan 0° - 5 \cot 270° + 3 \csc 90°$

6. $10 \csc 2\pi - 5 \cot 0$

7. $\tan \dfrac{3\pi}{2} \csc \pi - 5 \cot 2\pi$

8. $\cos 0 \sin \dfrac{\pi}{2} + 3 \tan 2\pi$

9. $\csc 180° \csc 0° + \cot 0°$

10. $8 \csc 0 \cot \dfrac{3\pi}{2} + 5 \sec \dfrac{\pi}{2}$

11. $\dfrac{4 \sin 0 \cos \pi}{\sin 3\pi/2}$

12. $2 \csc 270° \sec 0° - 4 \tan 180°$

13. $3 \cos 90° - 6 \sin 180° + \tan 90°$

14. $\sin \pi \cos \dfrac{\pi}{2} - \cos \dfrac{3\pi}{2} \sin 2\pi$

15. $5 \tan 270° \cot 360° - 3 \sec 270°$

16. $3 \sin 0° \cos 270° - 5 \cos 0°$

17. $\dfrac{5 \sec 0}{-3 \cos 2\pi} - 2 \tan 0$

18. $3 \sec 90° - 2 \csc 360° + 5 \cot 180°$

19. What is the numerical value of $\dfrac{\sin 360°}{\cos 360°} - \tan 360°$?

(A) 1 (B) 0 (C) −1 (D) −2 (E) undefined

20. What is the numerical value of $5 \csc \dfrac{3\pi}{2} \sec 2\pi - 4 \cot \dfrac{\pi}{2}$?

(A) 1 (B) 5 (C) −1 (D) −5 (E) undefined

Objective 2.2

Given the degree or radian measure of an angle, use a scientific calculator to find approximate values of the trigonometric functions for the angle.

Discussion

Scientific calculators are constructed to evaluate commonly used functions, including the sine, cosine, and tangent functions, with a single keystroke. The keys used to evaluate the trigonometric functions are typically labeled SIN , COS and TAN . With some calculators, it may be necessary to press a SHIFT or 2nd key to access these functions. Consult the operator's manual for the machine you are using.

Before a calculator can be used to evaluate a trigonometric function, it must be set to perform the calculations in the units being used to express the angles involved. If angles are in radians, the calculator must be set in radian mode. If angles are in degrees, it must be set in degree mode. Different calculators require different procedures in order to set the mode of angular measure. On a basic scientific calculator, the procedure usually involves a key labeled DRG (for Degree-Radian-Grad). On an advanced scientific or graphics calculator, the procedure usually involves selecting radians or degrees from a menu accessed by pressing a key labeled MODE . Consult the operator's manual for the calculator you are using in order to learn how to set the units of angular measure.

To evaluate a trigonometric function on a basic scientific calculator, enter the measure of the angle, including the sign. Then press the key corresponding to the function being evaluated. The value of the function, accurate to as many decimal places as the calculator displays, will appear in the display. To evaluate a trigonometric function on an advanced scientific or graphics calculator, first press the key corresponding to the function being evaluated. Then enter the measure of the angle, including the sign. The display will show the function being evaluated. Verify that you entered the function and angle correctly, and make corrections if necessary. Finally, press the key that directs the calculator to execute the evaluation. On most machines this key is labeled ENTER . The value of the function, accurate to as many decimal places as the calculator is set to display, will appear in the display.

Scientific calculators do not have special keys for the cosecant, secant and cotangent. These functions are evaluated by using their reciprocal relationships with the sine, cosine and tangent.

Recall that pairs of numbers of the form x and $\dfrac{1}{x}$ are called **reciprocals**. For example, $\dfrac{2}{5} = 0.4$ is the reciprocal of $\dfrac{5}{2} = 2.5$ and $\dfrac{5}{2} = 2.5$ is the reciprocal of $\dfrac{2}{5} = 0.4$. From the equations

$$\sin \theta = \frac{y}{r} \qquad \text{and} \qquad \csc \theta = \frac{r}{y},$$

which define the sine and cosecant, we see that these functions are reciprocals of each other. In the same way, we see that

$$\cos \theta = \frac{x}{r} \qquad \text{and} \qquad \sec \theta = \frac{r}{x},$$

and

$$\tan \theta = \frac{y}{x} \qquad \text{and} \qquad \cot \theta = \frac{x}{y}$$

are reciprocals. The following equations, called the **reciprocal formulas**, express the fact that these pairs of functions are reciprocals.

(1)

$$\csc \theta = \frac{1}{\sin \theta}$$

$$\sec \theta = \frac{1}{\cos \theta}$$

$$\cot \theta = \frac{1}{\tan \theta}$$

Most scientific calculators have a key, usually labeled $\boxed{1/x}$ or $\boxed{x^{-1}}$, for computing the reciprocal of a number with a single stroke. Enter a number x, press $\boxed{1/x}$ and the reciprocal of x (to as many decimal places as the calculator displays) appears. This key makes it easy to use the reciprocal formulas to evaluate the cosecant, secant and cotangent.

To evaluate the cosecant of an angle, first follow the steps outlined above to calculate the sine of this angle. According to the reciprocal formulas, the cosecant is just the reciprocal of the number calculated. Press the $\boxed{1/x}$ key to find the reciprocal of this number and hence the cosecant of the angle. Use a similar procedure to find the secant of an angle from the cosine.

When the tangent of an angle is defined the cotangent can be found from its tangent by using the reciprocal key. When the tangent is undefined, the cotangent is defined and has value 0. However, the reciprocal of the tangent is undefined, so the cotangent cannot be evaluated using the reciprocal key.

Summary

1. A scientific calculator can be used to evaluate the sine, cosine and tangent for a given angle. First set the calculator to the desired mode of angular measure (degrees or radians). Then follow the specific steps given in the operator's manual for the calculator to evaluate the sine, cosine and tangent of the angle. Failure to set the calculator to the desired mode of angular measure (degrees or radians) is the most common error in evaluating trigonometric functions.

2. Use the reciprocal relations between the cosecant and the sine, the secant and the cosine, and the tangent and the cotangent, together with the reciprocal key on a scientific calculator, to evaluate the cosecant, secant and cotangent for a given angle.

Examples

In the Examples which follow "BSC" is used to signify a basic scientific calculator and "ASGC" signifies an advanced scientific or graphics calculator. The ASGC used here is the TI-82®

Example 1. Find sin 50°. Round off the answer to four decimal places.

Solution 1 for the BSC.

Step 1. Select the degree mode.

Step 2. Enter the number 50

Step 3. Press the [SIN] key.

The answer then appears in the display:

$$\sin 50° = 0.76604444.$$

Step 4. Round off to four decimal places:

$$\sin 50° = 0.7660.$$

Solution 1 for the ASGC.

Step 1. Select the degree mode.

Step 2. Press the [SIN] key.

Step 3. Enter the number 50.

Step 4. Press [ENTER].

The answer then appears in the display:

$$\sin 50° = 0.7660444431.$$

Step 5. Round off to four decimal places:

$$\sin 50° = 0.7660.$$

The value of sin 50° cannot be given exactly by a number with a finite number of decimal places. Consequently, 0.7660, 0.76604444 and 0.7660444431 are all approximations to sin 50°. It would be more accurate to use the symbol " ≈ " in place of " = ". For the sake of simplicity, however, we will use the symbol " = " in our calculations.

Example 2. Find tan 15.7. Round off the answer to four decimal places.

Solution 2 for the BSC.

Step 1. Select the radian mode.

Step 2. Enter the number 15.7.

Solution 2 for the ASGC.

Step 1. Select the radian mode.

Step 2. Press the [TAN] key.

Step 3. Press the [TAN] key.

$$\tan 15.7 = -0.00796344$$

Step 4. Round off to four decimal places:

$$\tan 15.7 = -0.0080.$$

Step 3. Enter the number 15.7.

Step 4. Press [ENTER].

The answer then appears in the display:

$$\tan 15.7 = -0.0079634363.$$

Step 5. Round off to four decimal places:

$$\tan 15.7 = -0.0080.$$

Example 3. Find sec $(-129.8°)$. Round off the answer to four decimal places.

Solution 3 for both the BSC and the ASGC.

Since $\sec \theta = \dfrac{1}{\cos \theta}$, we first compute $\cos(-129.8°)$ as in Example 1.

Step 1. We obtain

$$\cos(-129.8°) = -0.64010970.$$

Step 2. Press the reciprocal key [1/x] (or [x⁻¹]) to obtain $\sec(-129.8°)$:

$$\sec(-129.8°) = -1.56223223.$$

Step 3. Round off the answer to four decimal places:

$$\sec(-129.8°) = -1.5622.$$

Example 4. Find $\cot \dfrac{\pi}{6}$. Round off the answer to four decimal places.

Solution 4 for the BSC.

Since $\cot \theta = \dfrac{1}{\tan \theta}$, begin by computing $\tan \dfrac{\pi}{6}$ as in Example 2.

Step 1. Select the radian mode.

Step 2. Enter the number $\dfrac{\pi}{6}$ by pressing the [π] key, the division key, the [6] key, and finally, the [=] key. The numerical value of $\dfrac{\pi}{6}$ expressed to as many decimal places as the calculator displays will appear in the display.

Step 3. Press [TAN].

The value of $\tan \dfrac{\pi}{6}$ then appears in the display:

$$\tan \dfrac{\pi}{6} = 0.57735027.$$

Solution 4 for the ASGC.

Step 1. Select the radian mode.

Step 2. Press [TAN], the left paren key, [2nd], [π], [÷], [6], and finally, the right paren key. The calculator display should show "tan(π/6)." The parentheses are needed to instruct the machine to divide π by 6 before applying the tangent function.

Step 3. Press [ENTER].

The value of $\tan \dfrac{\pi}{6}$ then appears in the display:

$$\tan \dfrac{\pi}{6} = 0.577350269.$$

Step 4. Press the reciprocal key which may be labeled

Step 4. Press the reciprocal key which may be labeled either $\boxed{1/x}$ or $\boxed{x^{-1}}$.

The answer then appears in the display:

$$\cot\frac{\pi}{6} = 1.73205081.$$

Step 5. Round off to four decimal places:

$$\cot\frac{\pi}{6} = 1.7321.$$

either $\boxed{1/x}$ or $\boxed{x^{-1}}$. Then press $\boxed{\text{ENTER}}$.
The answer then appears in the display:

$$\cot\frac{\pi}{6} = 1.73205081$$

Step 5. Round off to four decimal places:

$$\cot\frac{\pi}{6} = 1.7321.$$

Practice Problems

In **Problems 1 - 10**, find an approximate value for each function using a scientific calculator. Round off the answers to four decimal places.

1. $\sin 880°$

2. $\cos 8\pi$

3. $\tan(-450°)$ (Explain your answer.)

4. $\csc 0.7$

5. $\sec(-45°)$

6. $\cot 35$

7. $\tan(-236°)$

8. $\sin\dfrac{-\pi}{6}$

9. $\cot 60°$

10. $\cot \pi$ (Explain your answer.)

In **Problems 11 - 20**, find the value of the given expression. Round off the answer to four decimal places.

11. $14 \sin 52° \cos(-36°)$

12. $3 \csc\dfrac{3\pi}{4} - 2 \cos\dfrac{-5\pi}{4}$

13. $2 - \tan\dfrac{-3\pi}{7} \cot\dfrac{8\pi}{3}$

14. $\sec 213°\left[1 - \cot(-38°) \tan 116°\right]$

15. $\sin 22° \cos 108° - \sin 108° \cos 22°$

16. $\tan\dfrac{-9\pi}{5}\left[\cos\dfrac{3\pi}{5} + 2 \sec\dfrac{\pi}{6}\right]$

17. $\csc(-45°) \tan 30° - \sec(-30°) \sin 60°$

18. $4 \sin\dfrac{3\pi}{7}\left[1 - \csc\left(\dfrac{-7\pi}{3}\right) \tan\dfrac{8\pi}{5}\right]$

19. $\sec\left(\dfrac{-\pi}{3}\right)\left(3 \sin\dfrac{5\pi}{4} - 5 \cos\dfrac{\pi}{6}\right)$

20. $7 \sin(-27°) \cos(-213°) - \cot 151°$

21. What is the value of $\tan 97°\left[5 - \sin 76° \csc(-13°)\right]$, rounded to four decimal places?

(A) -116.5730 (B) -75.8513 (C) -36.4084 (D) 5.5922 (E) 9.3134

22. What is the value of $3 \sin\dfrac{2\pi}{13} - \cot\dfrac{7\pi}{6} \cos\dfrac{9\pi}{5}$, rounded to three decimal places?

(A) -15.510 (B) -0.940 (C) -0.007 (D) 1.197 (E) 15.560

50

Objective 2.3

Given an angle between $0°$ and $360°$ (0 and 2π radians), find its reference angle.

Discussion

The inverse problem is the problem of finding the angles for which the sine, cosine or tangent has a specified numerical value. The problem of finding all angles between $0°$ and $360°$ for which $\sin \theta = 0.821$ is a specific instance of the inverse problem. The idea of the reference angle is used to solve the inverse problem.

The **reference angle** θ' for an angle θ of any size in standard position is the smaller of the two positive angles formed by the terminal side of the angle and the x-axis. The reference angle is determined entirely by the position of the terminal side of the angle θ, so any two angles which have the same terminal side necessarily have the same reference angle.

When θ is a quadrant I angle of less than one revolution counterclockwise (so $0° \leq \theta < 360°$ or $0 \leq \theta < 2\pi$), the smaller of the two positive angles formed by the terminal side of the angle and the x-axis is just θ itself. Thus, θ is its own reference angle and $\theta' = \theta$. Figure 2.2 (a) illustrates this situation.

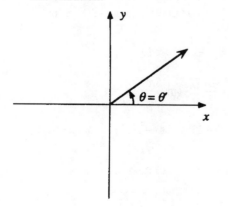

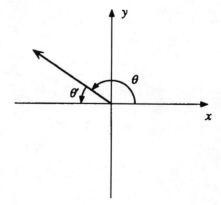

Figure 2.2 (a)
When $0 \leq \theta < 2\pi$ and θ is a quadrant I angle, $\theta' = \theta$.

Figure 2.2 (b)
When $0 \leq \theta < 2\pi$ and θ is a quadrant II angle, $\theta' = 180° - \theta$ (degrees) or $\theta' = \pi - \theta$ (radians).

When θ is a quadrant II angle and $0° \leq \theta < 360°$ (or $0 \leq \theta < 2\pi$), its reference angle has the terminal side of θ as its initial side and the negative x-axis as its terminal side. The negative x-axis is the terminal side of a $180°$ angle in standard position and θ is less than $180°$, so $\theta' = 180° - \theta$ (θ in degrees) or $\theta' = \pi - \theta$ (θ in radians). Figure 2.2 (b) illustrates this situation.

When θ is a quadrant III angle and $0° \leq \theta < 360°$ (or $0 \leq \theta < 2\pi$), its reference angle has the negative x-axis as its initial side and the terminal side of θ as its terminal side. The negative x-axis is the terminal side of a $180°$ angle in standard position and θ is greater than $180°$, so in this case $\theta' = \theta - 180°$ (θ in degrees) or $\theta' = \theta - \pi$ (θ in radians). Figure 2.2 (c) illustrates this situation.

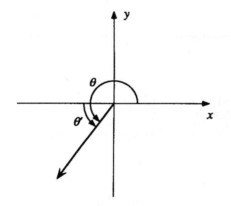

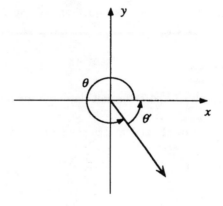

Figure 2.2 (c)
When $0 \leq \theta < 2\pi$ and is a quadrant III angle,
$\theta' = \theta - 180°$ (degrees)
or $\theta' = \theta - \pi$ (radians).

Figure 2.2 (d)
When $0 \leq \theta < 2\pi$ and is a quadrant IV angle,
$\theta' = 360° - \theta$ (degrees)
or $\theta' = 2\pi - \theta$ (radians).

When θ is a quadrant IV angle and $0° \leq \theta < 360°$ (or $0 \leq \theta < 2\pi$), its reference angle has the terminal side of θ as its initial side and the positive x-axis as its terminal side. In this case, the positive x-axis is the terminal side of a $360°$ angle in standard position and θ is less than $360°$, so $\theta' = 360° - \theta$ (θ in degrees) or $\theta' = 2\pi - \theta$ (θ in radians). Figure 2.2 (d) illustrates this situation.

The reference angle of a quadrantal angle with terminal side on the x-axis is $0°$ (0 radians). The reference angle of a quadrantal angle with terminal side on the y-axis is $90°$ $\left(\dfrac{\pi}{2} \text{ radians} \right)$.

When θ is not between $0°$ and $360°$ (0 and 2π), add or subtract revolutions (integer multiples of $360°$ or 2π) to obtain an angle between $0°$ and $360°$ (0 and 2π) that has the same terminal side as the original angle. The reference angle for this angle, which can be found as outlined above, is also the reference angle for θ.

Summary

1. The reference angle θ' for an angle θ in standard position is the smaller of the two positive angles formed by the terminal side of the angle and the x-axis. Two angles which have the same terminal side necessarily have the same reference angle.

2. To find the reference angle for an angle θ between $0°$ and $360°$ (0 and 2π radians) in standard position, sketch the angle in a coordinate system. Examine the sketch to see that
- when θ is a quadrant I angle, $\theta' = \theta$;
- when θ is a quadrant II angle, $\theta' = 180° - \theta$ (degrees) or $\theta' = \pi - \theta$ (radians);
- when θ is a quadrant III angle, $\theta' = \theta - 180°$ (degrees) or $\theta' = \theta - \pi$ (radians); and
- when θ is a quadrant IV angle, $\theta' = 360° - \theta$ (degrees) or $\theta' = 2\pi - \theta$ (radians).

Examples

In **Examples 1** and **2**, the calculation of the reference angle is illustrated.

Example 1. Draw the angle $\theta = 153°$ in standard position and find its reference angle.

Solution 1.
Since $153°$ is a quadrant II angle (see Figure 2.3), the reference angle is
$$\theta' = 180° - 153° = 27°.$$

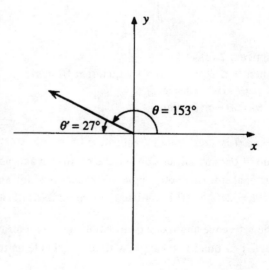

Figure 2.3 $\theta = 153°$, $\theta' = 27°$

Example 2. Draw the angle $\theta = \dfrac{5\pi}{4}$ in standard position and find its reference angle.

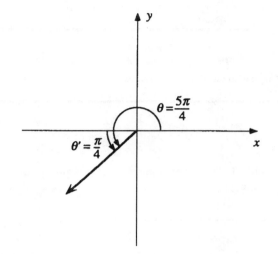

Solution 2.

Since $\dfrac{5\pi}{4}$ is a quadrant III angle, its reference angle is

$$\theta' = \frac{5\pi}{4} - \pi = \frac{\pi}{4}.$$

Figure 2.4 $\theta = \dfrac{5\pi}{4}, \ \theta' = \dfrac{\pi}{4}$

Practice Problems

In **Problems 1 - 12**, draw the given angle in standard position and find its reference angle.

1. $343°$

2. $\dfrac{7\pi}{8}$

3. $28°$

4. $\dfrac{7\pi}{6}$

5. $143.1°$

6. $\dfrac{3\pi}{2}$

7. $269°$

8. $\dfrac{5\pi}{3}$

9. $166°$

10. $\dfrac{11\pi}{12}$

11. $\dfrac{5\pi}{12}$

12. $180°$

13. What is the reference angle for $\theta = \dfrac{5\pi}{3}$?

(A) $\theta' = \dfrac{2\pi}{3}$ (B) $\theta' = \dfrac{\pi}{3}$ (C) $\theta' = \dfrac{\pi}{6}$ (D) $\theta' = \dfrac{\pi}{2}$ (E) $\theta' = \dfrac{\pi}{4}$

14. Which of the following angles have $\theta' = 13°$ as the reference angle? There is at least one correct response. Choose all correct responses.

(A) $\theta = 103°$ (B) $\theta = 193°$ (C) $\theta = 347°$ D) None of these

Objective 2.4

Given an angle between $0°$ and $360°$ (0 and 2π radians), express the values of the trigonometric functions for the given angle in terms of its reference angle.

Discussion

An angle θ and its reference angle θ' are closely related. The trigonometric functions have the same values for these two angles up to the $+$ or $-$ sign. That is to say,

(**2**)

$$
\begin{array}{ll}
\sin \theta = \pm \sin \theta', & \csc \theta = \pm \csc \theta', \\
\cos \theta = \pm \cos \theta', & \sec \theta = \pm \sec \theta', \\
\tan \theta = \pm \tan \theta', \text{ and} & \cot \theta = \pm \cot \theta'.
\end{array}
$$

To see this, consider a quadrant II angle θ as shown in Figure 2.5. Both θ and θ' are drawn in standard position. Points $P(x, y)$ and $P'(x', y')$ have been chosen on their terminal sides in such a way that $y = y'$.

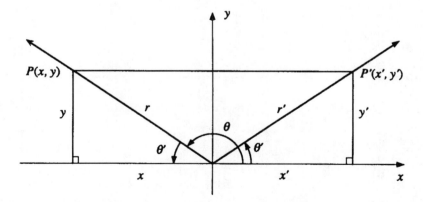

Figure 2.5

From the symmetry in the figure we see that

$$x = -x', \ y = y', \text{ and } r = r'.$$

Thus,

$$\sin \theta = \frac{y}{r} = \frac{y'}{r'} = \sin \theta', \qquad \csc \theta = \frac{r}{y} = \frac{r'}{y'} = \csc \theta',$$

$$\cos \theta = \frac{x}{r} = -\frac{x'}{r'} = -\cos \theta', \qquad \sec \theta = \frac{r}{x} = -\frac{r'}{x'} = -\sec \theta',$$

$$\tan \theta = \frac{y}{x} = -\frac{y'}{x'} = -\tan \theta', \qquad \cot \theta = \frac{x}{y} = -\frac{x'}{y'} = -\cot \theta'.$$

Table 2.2

Similar results hold for angles θ in the other three quadrants.

The proper choice of the $+$ or $-$ signs in equations (2) is determined by the quadrant of θ and the associated sign of the trigonometric function (see Figure 1.47). If θ is in a quadrant where the particular trigonometric function is positive, use the positive (+) sign. If θ is in a quadrant where the particular trigonometric function is negative, use the negative (−) sign. For example, for $\theta = 150°$ the reference angle is $\theta' = 30°$. Thus,

$$\sin 150° = \pm \sin 30° \text{ and } \cos 150° = \pm \cos 30°,$$

where we must make the proper choice of the $+$ or $-$ sign in each equation. We recognize that $150°$ is a quadrant II angle. In quadrant II the sine is positive and the cosine is negative. Consequently,

$$\sin 150° = +\sin 30° \text{ and } \cos 150° = -\cos 30°.$$

Summary

1. A reference angle is a quadrant I angle, a $0°$ (0 radian) quadrantal angle, or a $90°$ $\left(\dfrac{\pi}{2} \text{ radian}\right)$ quadrantal angle. Thus, for any reference angle θ' the values of the trigonometric functions are either non-negative or undefined.

2. Suppose angle θ has reference angle θ'. If θ lies in a quadrant where a trigonometric function is positive, then the trigonometric function has the same values for θ and θ'. If θ lies in a quadrant where a trigonometric function is negative, then the value of the trigonometric function for θ is the negative of the value of the function for θ'.

56

Examples

Example 1. Express $\tan 295°$ and $\csc 295°$ in terms of the reference angle.

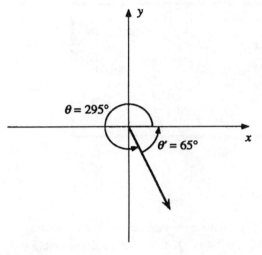

Solution 1.
First, $\theta = 295°$ is a quadrant IV angle and its reference angle is
$$\theta' = 360° - 295° = 65°.$$
Second, by equations (2), we have
$$\tan 295° = \pm\tan 65° \text{ and } \csc 295° = \pm\csc 65°,$$
where we must make a choice of signs.
Third, in quadrant IV the tangent is negative and the cosecant is negative, so
$$\tan 295° = -\tan 65° \text{ and } \csc 295° = -\csc 65°.$$
Verify these equalities with a calculator.

Figure 2.6

Example 2. Express $\cot \dfrac{2\pi}{3}$ and $\sin \dfrac{2\pi}{3}$ in terms of the reference angle.

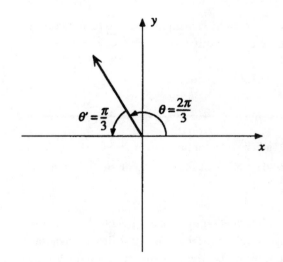

Solution 2.
First, $\theta = \dfrac{2\pi}{3}$ is a quadrant II angle and its reference angle is
$$\theta' = \pi - \frac{2\pi}{3} = \frac{\pi}{3}.$$
Second, by equations (2), we have
$$\cot\frac{2\pi}{3} = \pm\cot\frac{\pi}{3} \text{ and } \sin\frac{2\pi}{3} = \pm\sin\frac{\pi}{3},$$
where we must make a choice of signs.
Third, in quadrant II the cotangent is negative and the sine is positive, so
$$\cot\frac{2\pi}{3} = -\cot\frac{\pi}{3} \text{ and } \sin\frac{2\pi}{3} = +\sin\frac{\pi}{3}.$$
Verify these equalities with a calculator.

Figure 2.7

Practice Problems

In **Problems 1 - 24**, express the given value of the trigonometric function in terms of the reference angle.

1. $\tan 341°$

2. $\csc \dfrac{7\pi}{8}$

3. $\cos 28°$

4. $\sin \dfrac{7\pi}{6}$

5. $\cot 322°$

6. $\sec 324°$

7. $\csc 199.9°$

8. $\cot \dfrac{9\pi}{7}$

9. $\sin \dfrac{7\pi}{12}$

10. $\tan 265°$

11. $\cos \dfrac{11\pi}{12}$

12. $\sec 157°$

13. $\sin 173°$

14. $\tan \dfrac{\pi}{3}$

15. $\sec \dfrac{23\pi}{29}$

16. $\cos 199°$

17. $\csc \dfrac{2\pi}{7}$

18. $\cot 148°$

19. $\sin 359°$

20. $\tan \dfrac{4\pi}{7}$

21. $\sec \dfrac{3\pi}{7}$

22. $\cos \dfrac{13\pi}{15}$

23. $\csc 287°$

24. $\cot \dfrac{5\pi}{11}$

25. Which one of the following expresses $\tan 193°$ in terms of the reference angle for $\theta = 193°$?

(A) $-\tan(-13°)$ **(B)** $-\tan 167°$ **(C)** $\cot 77°$ **(D)** $\tan 13°$ **(E)** $-\cot(-77°)$

26. Let $\theta = \dfrac{7\pi}{12}$ and let θ' be the reference angle for θ . Which of the following three equations are correct? There is at least one correct response. Choose all correct responses.

(A) $\cos \theta = -\cos \theta'$ **(B)** $\sin \theta = -\sin \theta'$ **(C)** $\tan \theta = -\tan \theta'$ **(D)** None of these

---

Objective 2.5

(a) **Given a numerical value $a \geq 0$ for $\sin \theta$, $\cos \theta$, or $\tan \theta$, find all angles between $0°$ and $90°$ (0 and $\frac{\pi}{2}$ radians) which have the given function value, or state that there is no such angle.**
(b) **Given a numerical value for $\sin \theta$, $\cos \theta$, or $\tan \theta$, find all angles between $0°$ and $360°$ (0 and 2π radians) which have the given function value, or state that there is no such angle (*i.e.*, solve the inverse problem for the sine, cosine and tangent).**

Discussion

Suppose that a is a specific number of any sign or size. Is there necessarily an angle θ such that $\sin \theta = a$? ... such that $\cos \theta = a$? ... such that $\tan \theta = a$? If so, for what angles (between $0°$ and $360°$ or between 0 and 2π radians) do the functions have this value? This is the inverse problem for the sine, cosine and tangent.

First, consider the equations $\sin \theta = a$ and $\cos \theta = a$. The sine and cosine functions are defined by $\sin \theta = \frac{y}{r}$ and $\cos \theta = \frac{x}{r}$, where (x, y) is any point on the terminal side of θ and $r = \sqrt{x^2 + y^2}$ is the distance between this point and the origin. For any point (x, y) other than the origin, according to the Pythagorean Theorem $x^2 + y^2 = r^2$, so

$$(\cos \theta)^2 + (\sin \theta)^2 = \left(\frac{x}{r}\right)^2 + \left(\frac{y}{r}\right)^2 = \frac{x^2 + y^2}{r^2} = 1.$$

Thus, for every angle θ,

(3)
$$\boxed{(\cos \theta)^2 + (\sin \theta)^2 = 1.}$$

From equation (3) we see that $(\cos \theta)^2 \leq 1$ and $(\sin \theta)^2 \leq 1$ so for every angle θ

(4)
$$\boxed{-1 \leq \cos \theta \leq 1 \quad \text{and} \quad -1 \leq \sin \theta \leq 1.}$$

The inequalities (4) tell us that the cosine and sine of an angle are always between -1 and 1. Consequently, the equations $\sin \theta = a$ and $\cos \theta = a$ have no solution when a is greater than 1 or less than -1 (*i.e.*, when $|a| > 1$).

When $-1 < a < 1$, there are two angles θ of less than one counterclockwise revolution (so $0° \le \theta < 360°$ or $0 \le \theta < 2\pi$) for which $\sin \theta = a$ and two for which $\cos \theta = a$. To see why, think about the sine function. If an angle θ has a point (x, y) on its terminal side such that $\sin \theta = \dfrac{y}{r} = a$, then the angle ψ which has $(-x, y)$ on its terminal side also has $\sin \psi = \dfrac{y}{r} = a$. When $x \ne 0$, θ and ψ are different angles, so the equation $\sin \theta = a$ has two solutions. When $x = 0$, however, which happens when $a = 1$ or $a = -1$, θ and ψ are the same angle and the equation $\sin \theta = a$ has only one solution. Now, consider the cosine function. If an angle θ has a point (x, y) on its terminal side such that $\cos \theta = \dfrac{x}{r} = a$, then the angle ψ which has $(x, -y)$ on its terminal side also has $\cos \psi = \dfrac{x}{r} = a$. When $y \ne 0$, θ and ψ are different angles and give two solutions to $\cos \theta = a$. When $y = 0$, however, which happens when $a = 1$ or $a = -1$, θ and ψ are the same angle and we have only one solution for $\cos \theta = a$.

The situation is somewhat different with the equation $\tan \theta = a$. The tangent is defined by $\tan \theta = \dfrac{y}{x}$ where (x, y) is any point on the terminal side of θ. Since y and x can each have any numerical values, their quotient can have any value. Consequently, for each number a there is at least one angle θ (of less than one counterclockwise revolution) such that $\tan \theta = a$. In fact, there are always two such angles. If an angle θ has a point (x, y) on its terminal side such that $\tan \theta = \dfrac{y}{x} = a$, then the angle ψ which has $(-x, -y)$ on its terminal side also has $\tan \psi = \dfrac{-y}{-x} = a$. Thus, for **every** number a there are two angles θ of less than one counterclockwise revolution (so $0° \le \theta < 360°$ or $0 \le \theta < 2\pi$) for which $\tan \theta = a$.

We now know that the equations $\sin \theta = a$ and $\cos \theta = a$ $(-1 < a < 1)$ and $\tan \theta = a$ (a arbitrary) have two solutions and that the equations $\sin \theta = \pm 1$ and $\cos \theta = \pm 1$ have one solution between $0°$ and $360°$ (0 and 2π).

Now consider the problem described in part (a) of the Objective; the problem of finding all angles θ in quadrant I for which the sine, cosine or tangent has a given value $a \ge 0$.

From the discussion above, we know that when $a > 1$ there are no angles θ, so, in particular, no quadrant I angles for which $\sin \theta = a$ or $\cos \theta = a$. When $0 \le a \le 1$, the point $\left(\sqrt{1 - a^2}, a \right)$ lies on the terminal side of a quadrant I angle θ (or a quadrantal angle, if $a = 0$ or 1) such that

$$\sin \theta = \frac{y}{r} = \frac{a}{1} = a.$$

Similarly, the point $\left(a, \sqrt{1-a^2}\right)$ lies on the terminal side of a quadrant I angle ψ (or a quadrantal angle, if $a = 0$ or 1) such that

$$\cos \psi = \frac{x}{r} = \frac{a}{1} = a.$$

In fact, θ and ψ are the only angles between $0°$ and $90°$ (0 and $\frac{\pi}{2}$ radians) for which $\sin \theta = a$ and $\cos \psi = a$.

For each $a \geq 0$ there is exactly one angle θ between $0°$ and $90°$ (0 and $\frac{\pi}{2}$ radians) for which $\tan \theta = a$. It is the angle θ which has the point $(1, a)$ on its terminal side.

The quadrant I angles which satisfy the equations $\sin \theta = a$, $\cos \theta = a$ and $\tan \theta = a$, where $0 \leq a$, can be found quickly using the inverse trigonometric function keys on a scientific calculator. On most machines these keys are labeled SIN^{-1}, COS^{-1} and TAN^{-1}, and are accessed by pressing $\boxed{\text{2nd}}$ followed by $\boxed{\text{SIN}}$, $\boxed{\text{COS}}$ or $\boxed{\text{TAN}}$. On others, the inverse trigonometric functions are accessed by pressing $\boxed{\text{INV}}$ followed by $\boxed{\text{SIN}}$, $\boxed{\text{COS}}$ or $\boxed{\text{TAN}}$. Consult the operator's manual for the machine you are using.

To use a basic scientific calculator to solve one of the equations $\sin \theta = a$, $\cos \theta = a$ or $\tan \theta = a$, where $0 \leq a$, first set the desired mode of angular measure. Next, enter the number a. Finally, press the corresponding inverse function key. The value of θ, in the units of angular measure you chose, accurate to as many decimal places as the calculator displays, will appear in the display. The calculator will display an error indicator when the equation has no solution, as is the case with the equations $\sin \theta = a$ and $\cos \theta = a$, when $a > 1$.

To use an advanced scientific or graphics calculator to solve one of the equations $\sin \theta = a$, $\cos \theta = a$ or $\tan \theta = a$, where $0 \leq a$, first set the desired mode of angular measure. Next, press the corresponding inverse function key. Then enter the number a. The display will show the inverse function and the number a as you entered them. Verify that you entered everything as you intended and make corrections as necessary. Finally, press $\boxed{\text{ENTER}}$. The value of θ, in the units of angular measure you chose, accurate to as many decimal places as the calculator is set to display, will appear in the display. The calculator will display an error message when the equation has no solution, as is the case with the equations $\sin \theta = a$ and $\cos \theta = a$, when $a > 1$.

Now consider the more general problem described in part (b) of the Objective; the problem of finding all angles θ of less than one revolution for which the sine, cosine or tangent has a given value a, which may or may not be positive. We know from our investigations earlier in this discussion that there may be no, one or two such angles.

The three equations $\sin \theta = a$, $\cos \theta = a$ or $\tan \theta = a$ are solved similarly. Let's analyze the problem of finding the angles of less than one revolution for which $\tan \theta = a$. We know from above that there are two such angles. If $a > 0$, they are a quadrant I and a quadrant III angle. If $a = 0$, they are the quadrantal angles with terminal sides along the x-axis. If $a < 0$, they are a quadrant II and a quadrant IV angle.

Begin by using the methods from the discussion of part (a) of the Objective to find a quadrant I angle ψ such that $\tan \psi = |a|$. Recall from Objective 2.4 that if θ' is the reference angle for θ, then $\tan \theta = \pm \tan \theta'$. Therefore, if θ is any angle with reference angle ψ, either

$$\tan \theta = \tan \psi = |a| \quad \text{or} \quad \tan \theta = -\tan \psi = -|a|.$$

Consequently, for any angle θ with reference angle ψ,

$$\tan \theta = \pm a.$$

When $a \neq 0$, there is an angle of less than one revolution in each quadrant that has reference angle ψ. For two of them, the tangent function has value a and for the other two it has value $-a$. The two angles for which the tangent function has value a are the solution to the equation.

To find all angles θ of less than one revolution for which $\sin \theta = a$, $\cos \theta = a$ or $\tan \theta = a$ proceed as follows. First, determine how many solutions the equation has ($\sin \theta = a$ and $\cos \theta = a$ have no solutions when $|a| > 1$). Second, replace a with $|a|$ in the given equation. Solve this new equation to obtain an angle ψ. Finally, find the angles θ which have reference angle ψ and lie in a quadrant where the sign of a is the same as the sign of the trigonometric function involved. These angles are the solutions to the original equation.

When $|a| > 1$, the equations $\sin \theta = a$ and $\cos \theta = a$ have no solution. If the procedure described above is used to try to solve these equations with $|a| > 1$, the calculator will display an error message when the inverse sine or cosine key is accessed. Experiment with your calculator.

In the above procedure, the calculator's inverse trigonometric function operations are applied only to non-negative numbers $|a|$. These keys will also accept negative inputs and produce a solution to the trigonometric equation. It is easier, however, to find the second solution using the procedure given than it is from using a negative input.

In this Unit we are concerned with the inverse problem only for the sine, cosine and tangent. By using the reciprocal formulas, however, we can also solve the inverse problem for the cosecant, secant and cotangent. For example, suppose we want to solve the equation

$$\sec \theta = 2$$

for angles θ between $0°$ and $360°$. By using the reciprocal formulas (1) we can rewrite this equation as

$$\frac{1}{\cos \theta} = 2 \quad \text{or} \quad \cos \theta = \frac{1}{2}.$$

This equation can be solved by the procedure outlined above.

62

Summary

1. The equations $\sin \theta = a$ and $\cos \theta = a$ have two solutions between $0°$ and $360°$ (0 and 2π) when $|a| < 1$, one when $|a| = 1$ and none when $|a| > 1$. For every number a, the equation $\tan \theta = a$ has two solutions between $0°$ and $360°$ (0 and 2π).

2. To use a basic scientific calculator to solve $\sin \theta = a$, $\cos \theta = a$ or $\tan \theta = a$, where $a \geq 0$, for an angle θ between $0°$ and $90°$ $\left(0 \text{ and } \dfrac{\pi}{2}\right)$, follow these steps. First, set the mode of angular measure. Next, enter the number a. Finally, press the corresponding inverse function key. The calculator will display an error indicator when the equation does not have a solution.

3. To use an advanced scientific or graphics calculator to solve $\sin \theta = a$, $\cos \theta = a$ or $\tan \theta = a$, where $a \geq 0$, for an angle between $0°$ and $90°$ $\left(0 \text{ and } \dfrac{\pi}{2}\right)$, follow these steps. First, set the mode of angular measure. Next, press the corresponding inverse function key. Then enter the number a. Finally, press ENTER. The calculator will display an error message when the equation has no solution.

4. To solve $\sin \theta = a$, $\cos \theta = a$ or $\tan \theta = a$, first determine whether the equation has a solution. If the equation has a solution, solve the related equation $\sin \theta = |a|$, $\cos \theta = |a|$ or $\tan \theta = |a|$ and obtain an angle ψ between $0°$ and $90°$ $\left(0 \text{ and } \dfrac{\pi}{2}\right)$. The angle(s) θ which has reference angle ψ and lies in a quadrant where the sign of a is the same as the sign of the trigonometric function involved is the solution(s) to the original equation.

Examples

Examples 1 and 2 illustrate how to find only **one** angle θ. A more general problem is to find **all** angles θ between $0°$ and $360°$ (0 and 2π radians) for which $\sin \theta$, $\cos \theta$, or $\tan \theta$ has a given numerical value. Examples 3 - 6 illustrate how to solve this problem.

Example 1. Find the angle θ between $0°$ and $90°$ such that $\sin \theta = 0.5968$. Round off the answer to the nearest tenth of a degree.

Solution 1 for the BSC.

Step 1. Select the degree mode.

Step 2. Enter the number 0.5968.

Step 3. Press INV and SIN.

The angle θ expressed in degrees then appears in

Solution 1 for the ASGC.

Step 1. Select the degree mode.

Step 2. Press 2nd and [SIN⁻¹].

Step 3. Enter the number 0.5968.

Step 4. Press ENTER.

the display :
$$\theta = 36.641057°.$$

Step 4. Round to the nearest tenth of a degree:
$$\theta = 36.6°.$$

The angle θ expressed in degrees is then displayed:
$$\theta = 36.64105667°$$

Step 5. Round to the nearest tenth of a degree:
$$\theta = 36.6°.$$

Example 2. Find the angle θ between 0 and $\dfrac{\pi}{2}$ such that $\tan \theta = 2.7933$. Round off the answer to four decimal places.

Solution 2 for the BSC.

Step 1. Select the radian mode.

Step 2. Enter the number 2.7933.

Step 3. Press the ⎡TAN⎤ key.
$$\theta = 1.2270129 \,(\text{radians}).$$

Step 4. Round to four decimal places:
$$\theta = 1.2270.$$

Solution 2 for the ASGC.

Step 1. Select the radian mode.

Step 2. Press the ⎡TAN⎤ key.

Step 3. Enter the number 2.7933.

Step 4. Press ⎡ENTER⎤.
$$\theta = 1.227012856 \text{ (radians)}.$$

Step 5. Round to four decimal places:
$$\theta = 1.2270.$$

Example 3. Find all angles θ between $0°$ and $360°$ such that $\sin \theta = 0.5968$. Round off the answers to the nearest tenth of a degree.

Solution 3. for both the BSC and ASGC.

Step 1. There are two angles θ between $0°$ and $360°$ for which $\sin \theta = 0.5968$.
First, determine the quadrants of these two angles. Since $\sin \theta = 0.5968$ is positive, and $\sin \theta$ is positive only in quadrants I and II, the first angle, θ_1, is in quadrant I and the second angle, θ_2, is in quadrant II.

Step 2. The angles θ_1 and θ_2 have the same reference angle θ'.

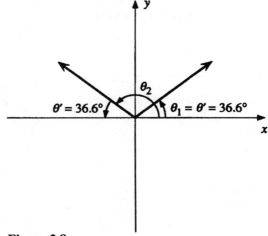

Figure 2.8

In order to find this reference angle, recall that in Objective 2.4 we learned that $\sin \theta = \pm\sin \theta'$ and that $\sin \theta'$ is positive. As a result,
$$\sin \theta' = |\sin \theta| = |0.5968| = 0.5968.$$
Now, from Example 1 above, we have $\theta' = 36.6°$.

Step 3. Draw the angles θ_1 and θ_2 in the correct quadrants with reference angle $\theta' = 36.6°$ as in Figure N2.8.

Step 4. Determine θ_1 and θ_2 from the quadrant and the reference angle. Use the drawing from Step 3 as a guide. We obtain

$$\theta_1 = \theta' = 36.6°$$

and

$$\theta_2 = 180° - \theta' = 180° - 36.6° = 143.4°.$$

Example 4. Find all angles θ between 0 and 2π such that $\tan \theta = -2.7933$. Round off the answers to four decimal places.

Solution 4. for both the BSC and ASGC.

Step 1. Determine the quadrants of the two angles θ_1 and θ_2.

Since $\tan \theta = -2.7933$ is negative and the tangent is negative in quadrants II and IV, θ_1 is in quadrant II and θ_2 is in quadrant IV.

Step 2. Find the reference angle for θ_1 and θ_2.

$$\tan \theta' = |-2.7933| = 2.7933.$$

From Example 2 above, we have

$$\theta' = 1.2270.$$

Step 3. Draw the angles θ_1 and θ_2 in the correct quadrants with reference angle $\theta' = 1.2270$ as in Figure 2.9.

Step 4. Determine θ_1 and θ_2 using the drawing from Step 3 as a guide. We obtain

$$\theta_1 = \pi - \theta' = \pi - 1.2270 = 1.9146$$

and

$$\theta_2 = 2\pi - \theta' = 2\pi - 1.2270 = 5.0562.$$

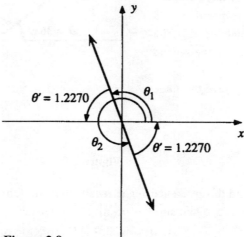

Figure 2.9

Example 5. Find all angles θ between $0°$ and $360°$ such that $\cos \theta = -0.5$. Round off the answers to the nearest tenth of a degree.

Solution 5.
Step 1. Determine the quadrants of θ_1 and θ_2.
Since $\cos \theta = -0.5$ is negative, θ_1 is in quadrant II and θ_2 is in quadrant III.
Step 2. Find the reference angle for θ_1 and θ_2.
Since $\cos \theta' = |-0.5| = 0.5$, $\theta' = 60°$.
Step 3. Draw angles θ_1 and θ_2 in the correct quadrants with reference angle $\theta' = 60°$ as in Figure 2.10.
Step 4. Determine θ_1 and θ_2.

$$\theta_1 = 180° - 60° = 120°$$

$$\theta_2 = 180° + 60° = 240°$$

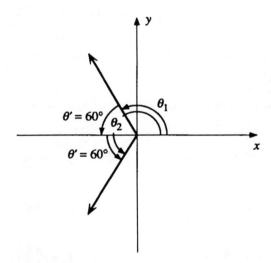

Figure 2.10

Example 6. Find all angles between 0 and 2π such that $\cos \theta = -1.2305$. Round off the answer to four decimal places.

Solution 6.
The cosine of an angle is always between -1 and 1. There are no angles θ such that $\cos \theta = -1.2305$.

Practice Problems

In **Problems 1 - 8**, find the quadrant I angle having the given function value. Use the indicated unit of angular measure. Round off the answers to two decimal places.

1. $\sin \theta = 0.5$ (degrees)
2. $\cos \theta = 0.3101$ (degrees)
3. $\tan \theta = 8$ (radians)
4. $\cos \theta = 0.75$ (radians)

5. $\tan \theta = 1.9$ (degrees)
6. $\sin \theta = 0.666$ (degrees)
7. $\cos \theta = 3$ (radians)
8. $\sin \theta = 2$ (radians)

In **Problems 9 - 16**, find the angles θ of less than one revolution having the given function value. Use the indicated unit of angular measure. Round off the answers to four decimal places.

9. $\sin \theta = -0.5$ (radians)
10. $\tan \theta = 0.83152$ (degrees)
11. $\cos \theta = 0.7071$ (radians)
12. $\tan \theta = -0.3710$ (degrees)

13. $\sin \theta = 0.7071$ (radians)
14. $\cos \theta = -0.92$ (degrees)
15. $\sin \theta = 1.5$ (radians)
16. $\cos \theta = -1.5$ (degrees)

17. Angles A and B are between $0°$ and $360°$, $\tan A = 4.7046$ and $\sin B = 0.2079$. Find the four possible values of $A + B$.

18. Angles α and β are between 0 and 2π, $\sin \alpha = -0.8660$ and $\cos \beta = 0.8660$. Find the four possible values of $\alpha + \beta$.

19. Angles P and Q are between 0 and 2π, $\cos P = -0.9239$ and $\tan Q = -0.7265$. Find the possible values of $P + Q$.

20. Angles θ and ϕ are between $0°$ and $360°$, $\cos \theta = -0.3907$ and $\tan \phi = -14.3007$. Find the possible values of $\theta + \phi$.

21. Angles C and D are between $0°$ and $360°$, $\cos C = 0.1392$ and $\sin D = -0.1908$. Which of the following are possible (approximate) values of $C + D$? There is at least one correct response. Choose all correct responses.

(A) $71°$ (B) $93°$ (C) $267°$ (D) $273°$ (E) $627°$

22. Angles θ and ϕ are between 0 and 2π, $\tan \theta = 0.0777$ and $\sin \phi = 0.5878$. Which of the following are possible (approximate) values for $\theta + \phi$? There is at least one correct response. Choose all correct responses.

(A) 0.6655 (B) 0.7059 (C) 2.9581 (D) 3.8475 (E) 5.7324

UNIT 2
Sample Examination 1

1. What is the value of $6 \cos(570°) \sec(170°) + 5 \tan(-230°)$ rounded to three decimal places?

(A) -0.682 **(B)** -4.004 **(C)** -5.196 **(D)** -6.388 **(E)** -6.825

2. Which of the following angles has $\theta' = \dfrac{5\pi}{16}$ as its reference angle? There is at least one correct response. Choose all correct responses.

(A) $\dfrac{3\pi}{16}$ **(B)** $\dfrac{11\pi}{16}$ **(C)** $\dfrac{21\pi}{16}$ **(D)** None of these

3. Let $\theta = 296°$ and let θ' be the reference angle for θ. Which of the following equations are true? There is at least one correct response. Choose all correct responses.

(A) $\csc \theta = \csc \theta'$ **(B)** $\tan \theta = -\tan \theta'$ **(C)** $\sec \theta = \sec \theta'$ **(D)** None of these

4. Angles α and β are between $0°$ and $360°$, $\sin \alpha = -0.7771$ and $\cos \beta = -0.9659$. Which of the following are possible (approximate) values for $\alpha + \beta$? There is at least one correct response. Choose all correct responses.

(A) $144°$ **(B)** $294°$ **(C)** $396°$ **(D)** $474°$ **(E)** $654°$

5. Angles ϕ and ψ are between 0 and 2π radians, $\sin \phi = -0.3466$ and $\tan \psi = 0.3374$. Which of the following are possible (approximate) values for $\phi + \psi$? There is at least one correct response. Choose all correct responses.

(A) 0.679 **(B)** 3.821 **(C)** 6.312 **(D)** 8.745 **(E)** 9.453

UNIT 2
Sample Examination 2

1. What is the value of $7 \tan \left(\frac{8\pi}{11}\right) \cot (-7.4) + \cos \left(\frac{29\pi}{5}\right)$ rounded to three decimal places?

(**A**) 12.413   (**B**) 4.751   (**C**) −1.201   (**D**) −3.113   (**E**) −8.115

2. Which of the following expresses $\tan 295°$ in terms of the reference angle for $\theta = 295°$?

(**A**) $-\cot 25°$ (**B**) $-\csc 28°$ (**C**) $-\sec 62°$ (**D**) $-\tan 65°$ (**E**) None of these

3. Let $\theta = \frac{11\pi}{7}$ and let θ' be the reference angle for θ. Which of the following equations are true? There is at least one correct response. Choose all correct responses.

(**A**) $\csc \theta = -\csc \theta'$ (**B**) $\tan \theta = -\tan \theta'$ (**C**) $\sec \theta = \sec \theta'$ (**D**) None of these

4. Angles ϕ and ψ are between 0 and 2π radians, $\sin \phi = -0.1709$ and $\cos \psi = 0.2821$. Which of the following are possible (approximate) values for $\phi + \psi$? There is at least one correct response. Choose all correct responses.

(**A**) 4.598   (**B**) 5.170   (**C**) 7.396   (**D**) 8.312   (**E**) 11.110

5. Angles R and S are between $0°$ and $360°$, $\cos R = -0.9877$ and $\tan S = -0.2126$. Which of the following are possible (approximate) values for $R + S$? There is at least one correct response. Choose all correct responses.

(**A**) 159°   (**B**) 177°   (**C**) 201°   (**D**) 363°   (**E**) 519°

UNIT 3

RIGHT TRIANGLES AND THE LAW OF SINES

Introduction

In this Unit we will use the trigonometric functions to solve right triangles. We also will begin a study of solving oblique triangles by introducing the Law of Sines. The **ambiguous case** of the Law of Sines, where there can be more than one triangle with specified sides and angle, will be examined in detail in this Unit. Our study of solving oblique triangles will be completed in Unit 4.

UNIT 3
RIGHT TRIANGLES AND THE LAW OF SINES

Objective 3.1

    (a) Given one side and one acute angle of a right triangle, solve the triangle.

    (b) Given two sides of a right triangle, solve the triangle.

Objective 3.2

Given one side and two angles of a (possibly oblique) triangle, solve the triangle using the Law of Sines.

Objective 3.3

Given two sides and an opposite angle of a (possibly oblique) triangle, solve the triangle using the Law of Sines (ambiguous case).

RIGHT TRIANGLES AND THE LAW OF SINES

Objective 3.1

    (a) **Given one side and one acute angle of a right triangle, solve the triangle.**

    (b) **Given two sides of a right triangle, solve the triangle.**

Discussion

One of the most important applications of trigonometry is solving problems involving triangles. A triangle consists of three sides and three angles. These six parts are interrelated and, with one exception, a triangle is completely determined by one side and two additional parts. To solve a triangle means to find the lengths of its sides and the measures of its angles from a knowledge of three of its six parts (including at least one side). When we know we are working with a right triangle, we know one of the angles measures 90°, so only two additional pieces of information are needed to solve the triangle.

Figure 3.1 shows a right triangle with vertices *A, B* and *C* labeled in the standard way. It is customary to label each vertex of a triangle with a capital letter and the side opposite a vertex with the same letter in lower case. The triangle is then designated by the word *triangle*, or the symbol Δ, followed by a list of the vertices. In right triangles, the vertex with the right angle is usually listed second. Thus, the right triangle in Figure 3.1 might be designated ΔACB or ΔBCA (or, less frequently, ΔABC).

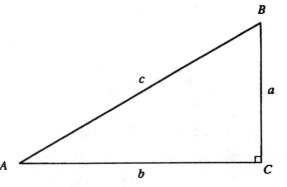

Figure 3.1

In a right triangle, the side opposite the right angle is called the **hypotenuse** of the triangle. Side c is the hypotenuse of $\triangle ACB$ shown in Figure N3.1. The other two sides of a right triangle are called **legs**.

There is little need to distinguish between an angle, the measure of the angle and the vertex where the angle appears in a triangle. As a result, we use the letters which label the vertices of a triangle to stand for all three and rely on the context to tell us which is intended.

Angles in a triangle are always viewed as positive angles formed by rotating a side through the interior of the triangle to the adjacent side. Angles of a triangle are almost always given in degrees. The sum of the measures of the three angles of a triangle which lies in a plane is $180°$. Thus, in any plane triangle ABC, $A + B + C = 180°$. It follows that every angle in a triangle is between $0°$ and $180°$.

The fact that one of the angles in a right triangle is a $90°$ angle makes solving them easier than solving oblique triangles. In a right triangle ACB with right angle C,

(1)
$$A + B = 90°.$$

and, according to the Pythagorean Theorem, the sum of the squares of the legs is equal to the square of the hypotenuse, so

(2)
$$a^2 + b^2 = c^2.$$

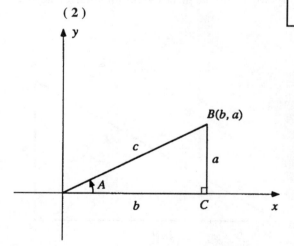

Equation (1) relates the acute angles of a right triangle. Equation (2) relates its sides. The trigonometric functions give equations that relate sides with angles. To illustrate this, draw a rectangular coordinate system on the triangle ACB from Figure 3.1 so the angle A is in standard position as shown in Figure 3.2. In this coordinate system the point B is on the terminal side of angle A and has coordinates (b, a). Consequently, from the definitions of the trigonometric functions,

Figure 3.2

(3)
$$\sin A = \frac{a}{c} = \frac{\text{side opposite } A}{\text{hypotenuse}},$$

(4)
$$\cos A = \frac{b}{c} = \frac{\text{side adjacent to } A}{\text{hypotenuse}},$$

(5)
$$\tan A = \frac{a}{b} = \frac{\text{side opposite } A}{\text{side adjacent to } A}.$$

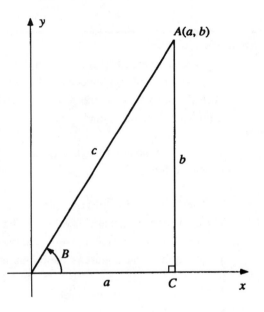

To obtain analogous equations relating angle B with the sides of the triangle, reflect $\triangle ACB$ through any of its sides and rotate the resulting triangle so it is positioned as in Figure 3.3. Now, apply the analysis used to obtain the equations involving angle A to obtain the equations involving angle B. Equations (6), (7) and (8) result:

Figure 3.3

(6)
$$\sin B = \frac{b}{c} = \frac{\text{side opposite } B}{\text{hypotenuse}},$$

(7)
$$\cos B = \frac{a}{c} = \frac{\text{side adjacent to } B}{\text{hypotenuse}},$$

(8)
$$\tan B = \frac{b}{a} = \frac{\text{side opposite } B}{\text{side adjacent to } B}.$$

74

By using equations (1) through (8) we can find the lengths of the sides and the measures of the angles of a right triangle from either the lengths of two sides or from the length of one side and the measure of one acute angle. If two sides of the triangle are known, find the remaining side from the Pythagorean Theorem (2), find the acute angle A (or B) by solving one of equations (3) - (5), and find the other acute angle B (or A) from equation (1). If the length of one side and the measure of one acute angle are given, find the remaining sides by using two of the equations from (3) - (8) that involve both the known angle and the known side. Find the other acute angle from equation (2). The Examples illustrate these methods.

Summary

1. It is customary to label each vertex of a triangle with a capital letter and the side opposite a vertex with the same letter in lower case. The triangle is then designated by the word *triangle*, or the symbol Δ, followed by a list of the vertices, *e.g.*, ΔACB. In a right triangle the 90° angle is usually labeled C and listed as the second vertex.

2. The side opposite the right angle in a right triangle is called the **hypotenuse**. The remaining sides are called **legs**.

3. The sum of the angles of a plane triangle ABC is 180°: $A + B + C = 180°$.
The sum of the acute angles A and B of a right triangle is 90°: $A + B = 90°$.
The sides of a right triangle are related by the Pythagorean Theorem. The sum of the squares of the legs is equal to the square of the hypotenuse: $a^2 + b^2 = c^2$.

4. Trigonometric functions relate the sides with the angles of a right triangle. For any acute angle θ of a right triangle,

$$\sin \theta = \frac{\text{side opposite } \theta}{\text{hypotenuse}}, \quad \cos \theta = \frac{\text{side adjacent to } \theta}{\text{hypotenuse}} \quad \text{and} \quad \tan \theta = \frac{\text{side opposite } \theta}{\text{side adjacent to } \theta}.$$

5. All the sides and all the angles of a right triangle can be found from two sides or from one side and one acute angle. If two sides of a right triangle are known, find the remaining side by using the Pythagorean Theorem. Use trigonometric equations relating sides and angles to find the acute angles. If one side and one acute angle are known, find the other sides by using the two trigonometric equations that involve the known angle, the known side and the unknown side. Find the second acute angle from the equation $A + B = 90°$.

Examples

Examples 1 - 3 below, illustrate the procedure for solving right triangles. In Examples 1 and 2, one side and one acute angle are given. In Example 3, two sides are given. *An accurate sketch is the secret to successfully solving triangles.*

Example 1. Solve right triangle ACB, where $a = 8$, $A = 35°$ and $C = 90°$.

Solution 1.
Step 1. Sketch the right triangle and label the known and unknown quantities.
Step 2. Solve for angle B by using

$$A + B = 90°.$$
$$35° + B = 90°$$
$$B = 90° - 35° = 55°$$

Step 3. Solve for side b by using a trigonometric ratio involving b, the side $a = 8$, and the angle $A = 35°$.

$$\tan 35° = \frac{8}{b}, \quad b = \frac{8}{\tan 35°} = 11.43.$$

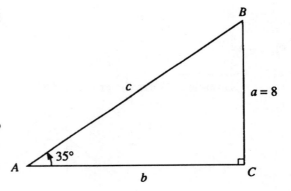

Figure 3.4

Step 4. Solve for side c by using a trigonometric ratio involving c, $a = 8$ and $A = 35°$.

$$\sin 35° = \frac{8}{c}, \quad c = \frac{8}{\sin 35°} = 13.95.$$

The complete solution of this right triangle is

$$a = 8 \qquad b = 11.43, \qquad c = 13.95,$$
$$A = 35°, \qquad B = 55°, \qquad C = 90°.$$

Note that Step 2 and Step 3 can be interchanged, and Step 4 can also be accomplished by using the Pythagorean Theorem.

Example 2. Find side b in right triangle ACB, where $c = 127.9$, $A = 10.1°$ and $C = 90°$.

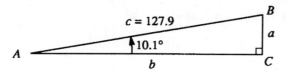

Solution 2.
Step 1. Sketch the right triangle and label the known and unknown quantities.
Step 2. Solve for side b by using a trigonometric ratio involving b and the given quantities, $c = 127.9$ and $A = 10.1°$.

$$\cos 10.1° = \frac{b}{127.9}$$
$$b = 127.9 \cos 10.1° = 125.92.$$

Figure 3.5

Example 3. Solve right triangle ABC, where $b = 89.6$, $c = 121.1$ and $C = 90°$.

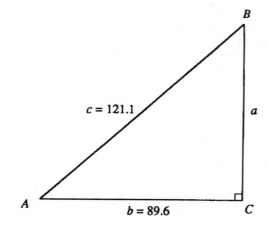

Solution 3.
Step 1. Sketch the right triangle and label the known and unknown quantities.
Step 2. Solve for angle A by using a trigonometric ratio.

$$\cos A = \frac{b}{c} = \frac{89.6}{121.1} = 0.7399 \text{ (see Objective 2.5)}$$
$$A = 42.3°$$

Step 3. Solve for angle B by using
$$A + B = 90°.$$
$$42.3° + B = 90°$$
$$B = 90° - 42.3° = 47.7°$$

Figure 3.6

Step 4. Solve for side a by using $a^2 + b^2 = c^2$.
$$a^2 + 89.6^2 = 121.1^2$$
$$a^2 = 121.1^2 - 89.6^2$$
$$a = 81.47$$

The complete solution of the right triangle is

$$a = 81.47, \qquad b = 89.6, \qquad c = 121.1,$$
$$A = 42.3°, \qquad B = 47.7°, \qquad C = 90°.$$

Note that in Step 4 we can use the trigonometric ratio $\sin A = \dfrac{a}{c}$ instead of the Pythagorean Theorem to find a. Using $A = 42.3°$ from Step 2 and $c = 121.1$, we obtain

$$\sin 42.3° = \frac{a}{121.1} \quad \text{and} \quad a = 121.1 \sin 42.3° = 81.5.$$

We obtain a slightly different value for a because of rounding off in the calculation of A in Step 2. By using the more accurate value $42.28°$ for A, we obtain $a = 81.47$.

Authors' Note

Pythagoras of Samos was a contemporary of Confucius and Buddha. No recorded scholarly work of his has survived nor has any one of the biographies which were written, including one by Aristotle. He was widely traveled, a prophet, a mystic and a mathematician. The phrase "All is number" is attributed to him. He established a society based in mathematical and philosophical reasoning. Because this society was secret and communal, it is difficult to trace certain bodies of work, discoveries or specific knowledge to a single individual. It is best we think in terms of the contributions of the Pythagoreans.

Practice Problems

1. Find c in the triangle shown in Figure 3.7.

2. Solve the right triangle shown in Figure 3.8.

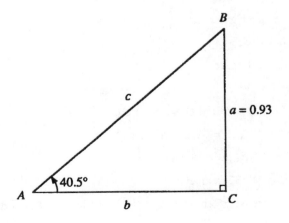

Figure 3.7

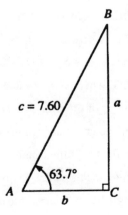

Figure 3.8

3. Find the length of side a in the triangle shown in Figure 3.9.

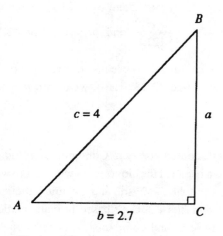

Figure 3.9

4. Solve the right triangle ACB, where $b = 18.3$, $B = 52°$ and $C = 90°$.

5. Solve the right triangle ACB, where $b = 170.5$, $c = 231$ and $C = 90°$.

6. Solve the right triangle ACB, where $A = 35°$, $c = 0.5$ and $C = 90°$.

7. Solve $\triangle ACB$, where $A = 17.3°$, $a = 17.8$ and $C = 90°$.

8. Solve $\triangle ACB$, where $a = 83.5$, $b = 17.2$ and $C = 90°$.

9. Solve $\triangle RST$, where $r = 3.6$, $T = 47°$ and S is a right angle.

10. Solve $\triangle MBL$, where $m = 0.13$, $l = 0.70$ and B is a right angle.

11. Solve $\triangle NBC$, where $n = 69$, $b = 101$ and B is a right angle.

12. Solve $\triangle CBS$, where $S = 31°$, $b = 66$ and B is a right angle.

13. Solve $\triangle REG$, where $R = 76.8°$, $r = 52.3$ and $E = 90°$.

14. Find the angles of the triangle PDQ, where $p = 12$, $q = 5$ and D is a right angle.

15. Find the perimeter of the triangle CSU, where $C = 43°$, $c = 89$ and S is a right angle.

16. Find the perimeter of the triangle UNM, where $U = 16°$, $n = 16$ and $N = 90°$.

17. Find the acute angles of $\triangle DRG$, where $r = 208$, $g = 132$ and $R = 90°$.

18. A right triangle has an acute angle of $22.7°$ and hypotenuse of length 48.2. What is the length of the side opposite the $22.7°$ angle?

19. A right triangle has a side of length 1.73 and hypotenuse of length 3.46. Find the acute angle opposite the side of length 1.73.

20. A right triangle has one acute angle of 83°. The leg of the triangle adjacent to this angle has length 2.5. What is the length of the hypotenuse?

21. What is the (approximate) perimeter of right triangle ACB, where $B = 9°$, $c = 126$ and $C = 90°$?

(A) 261   (B) 265   (C) 268   (D) 270   (E) 273

22. In $\triangle ACB$, $a = 56$, $b = 72$ and $C = 90°$. What is $A - B$?

(A) −14°   (B) −16°   (C) −18°   (D) −20°   (E) −21°

23. What is the (approximate) length of side a of right triangle CAR, where $c = 6.4$, $r = 8.7$ and $A = 90°$?

(A) 9.9   (B) 10.3   (C) 10.8   (D) 11.4   (E) 11.9

Objective 3.2

Given one side and two angles of a (possibly oblique) triangle, solve the triangle using the Law of Sines.

Discussion

A triangle which does not have a right (90°) angle is called an **oblique** triangle. As with all plane triangles, the sum of the angles of an oblique triangle is 180°. An oblique triangle can have no angles greater than 180° and no more than one angle greater than 90°, for in either circumstance the sum of the angles would exceed 180°. Thus, an oblique triangle either has three acute angles (each less than 90°) or it has two acute angles and one obtuse angle (greater than 90°). Figures 3.10 (a) and 3.10 (b) illustrate these possibilities. These figures also show the standard labeling of triangles. The angles (or vertices) are designated by capital letters and the side opposite an angle is designated by the same letter in lower case. This standard labeling is assumed in the statements of trigonometric relationships, including the Law of Sines and the Law of Cosines.

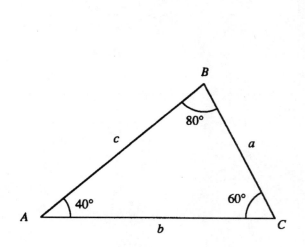

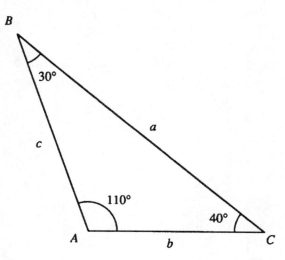

Figure 3.10 (a)   **Figure 3.10 (b)**

The familiar relationship $A + B + C = 180°$ among the angles of a triangle holds for oblique triangles as well as for right triangles. The Pythagorean Theorem and the equations relating sides and angles discussed in Objective 3.1, however, are true only for right triangles. They do not hold for oblique triangles. Our problem now is to discover general relationships among sides and among sides and angles which are true for both right and oblique triangles — relationships that can be used to solve any triangle, right or oblique, for which three parts (including a side) are known. The relationships we seek are called the **Law of Sines** and the **Law of Cosines**.

There are four possible situations that can occur when three parts of a triangle, including one side, are given.

Case I: One side and two angles are given.

Case II: Two sides and an angle opposite one of these sides are given.

Case III: Two sides and the angle included between these sides are given.

Case IV: Three sides are given.

The Law of Sines is used in Cases I and II. In the remainder of this Unit we develop the Law of Sines and apply it to these two Cases. In Unit 4 we will develop the Law of Cosines and apply it to Cases III and IV.

To discover the Law of Sines, consider an oblique triangle ABC with three acute angles as shown in Figure 3.11. Notice that the angles and sides are labeled in the standard way. From vertex B construct a perpendicular BD to the line containing AC. Let h denote the length of BD (h is the height of the triangle). This construction produces two right triangles ADB and CDB.

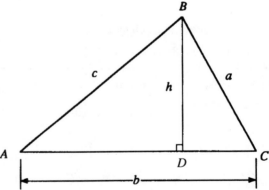

Figure 3.11

In right triangle ADB we have

$$\frac{h}{c} = \sin A \quad \text{so} \quad h = c \sin A.$$

In right triangle CDB we have

$$\frac{h}{a} = \sin C \quad \text{so} \quad h = a \sin C.$$

These two expressions for h must be equal, so

$$a \sin C = c \sin A.$$

On dividing this equation by $\sin A \, \sin C$, we finally obtain

$$\frac{a}{\sin A} = \frac{c}{\sin C}.$$

By constructing the perpendicular from vertex A, we obtain, by similar reasoning, the equation

$$\frac{b}{\sin B} = \frac{c}{\sin C}.$$

Thus, when ABC is an oblique triangle with three acute angles,

$$\frac{a}{\sin A} = \frac{b}{\sin B} = \frac{c}{\sin C}.$$

A slight modification of this reasoning leads to the conclusion that these equations also hold when ABC is an oblique triangle with an obtuse angle. When ABC is a right triangle with $C = 90°$, $\sin C = 1$ and these equations become

$$\frac{a}{\sin A} = \frac{b}{\sin B} = \frac{c}{1}.$$

These equations simply say that $\sin A = \dfrac{a}{c}$ and $\sin B = \dfrac{b}{c}$. So, the equations hold for triangles of every shape and size.

THEOREM   **Law of Sines:**

For any triangle ABC,

$$\frac{a}{\sin A} = \frac{b}{\sin B} = \frac{c}{\sin C},$$

where a, b and c are the lengths of the sides opposite the angles A, B and C.

Now suppose that one side, say side a, and two angles of a triangle are given. Since the sum of the angles must be $180°$, we can calculate the third angle immediately. Knowing a, A, B and C, we can calculate b and c by using the equations

$$\frac{b}{\sin B} = \frac{a}{\sin A} \quad \text{and} \quad \frac{c}{\sin C} = \frac{a}{\sin A},$$

from The Law of Sines, to obtain

$$b = \sin B \, \frac{a}{\sin A} \quad \text{and} \quad c = \sin C \, \frac{a}{\sin A}.$$

Summary

1. The equation $A + B + C = 180°$ holds for all triangles. The Pythagorean Theorem and the trigonometric ratios relating sides and angles, however, hold only for right triangles and cannot be used to solve oblique triangles. The Law of Sines and the Law of Cosines express relationships among sides and angles which are true for all triangles and can be used to solve triangles for which three parts (including a side) are known.

2.

> **The Law of Sines:**
>
> For any triangle ABC,
>
> $$\frac{a}{\sin A} = \frac{b}{\sin B} = \frac{c}{\sin C},$$
>
> where a, b and c are the lengths of the sides opposite the angles A, B and C.

3. To solve a (possibly oblique) triangle given one side and two angles, use the equation $A + B + C = 180°$ to find the third angle and the Law of Sines to find the two unknown sides.

Authors' Note

The Hindus, who seem to have acquired their knowledge of trigonometry from the Greeks, significantly transformed the study. Trigonometry from the time of Ptolemy was based on the relationship between the chords of a circle and the central angles they subtend. The writers of the *Siddhāntas* recorded their use of the length of the half-chord and their study of the "half-angle" subtended. This function was designated by the Hindu word *jya-ardha* (literally half-chord) and was abbreviated as *jya*.

In early Arabic astronomical calculations there was competition between the use of the Greek geometry of chords and the Hindu tables of half-chords. Ultimately, most Arabic trigonometry was built on the Hindu tables. In the Arabic use of the Hindu word *jya*, it phonetically became *jiba* which was abbreviated *jb*. Over time, because the word *jiba* has no Arabic meaning, the abbreviation *jb* became *jaib*, meaning curve, cavity or cove. During the 12th century, when Arabic works were being translated into Latin, the Latin equivalent to *jaib*, the word *sinus*, was used and became the word *sine*.

Thus was born, apparently in India and through a number of interesting etymological derivations, the predecessor of the modern trigonometric function known as the sine of an angle.

Examples

Example 1. Solve triangle ABC, where $a = 12$, $B = 110°$ and $C = 45°$.

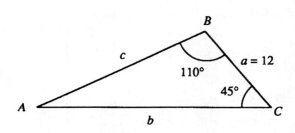

Solution 1.

Step 1. Sketch the triangle and label the known and unknown quantities.

Step 2. Solve for the third angle by using
$$A + B + C = 180°.$$
$$A + 110° + 45° = 180°$$
$$A = 180° - 110° - 45° = 25°$$

Figure 3.12

Step 3. Solve for side b by using the Law of Sines.

$$\frac{b}{\sin B} = \frac{a}{\sin A}$$

$$\frac{b}{\sin 110°} = \frac{12}{\sin 25°}$$

$$b = \sin 110° \cdot \frac{12}{\sin 25°} = 26.68$$

Step 4. Solve for side c by using the Law of Sines.

$$\frac{c}{\sin C} = \frac{a}{\sin A}$$

$$\frac{c}{\sin 45°} = \frac{12}{\sin 25°}$$

$$c = \sin 45° \cdot \frac{12}{\sin 25°} = 20.08$$

The complete solution of the triangle is

$$a = 12, \qquad b = 26.68, \qquad c = 20.08,$$
$$A = 25°, \qquad B = 110°, \qquad C = 45°.$$

Practice Problems

1. Find side a in the triangle shown in Figure 3.13.

2. Find side b in the triangle shown in Figure 3.14.

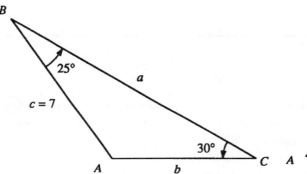

Figure 3.13   **Figure 3.14**

3. Find side b in triangle ABC, where $a = 11$, $A = 80°$ and $B = 65°$.

4. Find side c in triangle ABC, where $b = 81$, $A = 51°$ and $C = 34.5°$.

5. Find side a in triangle ABC, where $c = 175$, $B = 103.4°$ and $C = 37.3°$.

6. Solve triangle PQR, where $r = 72.1$, $P = 27°$ and $R = 25°$.

7. Solve triangle KMN, where $k = 0.27$, $M = 63°$ and $N = 70.1°$.

8. Solve triangle DAF, where $a = 43.9$, $D = 36.1°$ and $F = 60.9°$.

9. To the nearest tenth, what is the length of side a of (possibly oblique) triangle ABC, where $B = 13°$, $C = 56°$ and $b = 7.4$? Choose all correct responses.

(A) 27.3   **(B)** 27.9   **(C)** 28.3   **(D)** 30.2   **(E)** 30.7

10. To the nearest unit, what is the length of side c of (possibly oblique) triangle ABC, where $A = 49°$, $B = 63°$ and $a = 23$? Choose all correct responses.

(A) 28   **(B)** 29   **(C)** 30   **(D)** 31   **(E)** 32

Objective 3.3

Given two sides and an opposite angle of a (possibly oblique) triangle, solve the triangle using the Law of Sines (ambiguous case).

Discussion

Solving an oblique triangle in Case II (where two sides and an opposite angle are known) is more complicated than in Case I because it is possible to have no solution, one solution or two solutions. For this reason, Case II is called the **ambiguous case.**

Let us see geometrically how the possibilities of no solution, one solution or two solutions arise in Case II. Suppose we know two sides, a and c, and the angle A opposite side a. To draw a triangle ABC having these sides and this angle, first draw angle A as in Figure 3.15. Second, on the terminal side of A, measure off the length of side c as in Figure 3.16. This locates vertex B of the triangle. Finally, we must locate vertex C on the initial side of angle A. The distance between vertex B and vertex C must be the length of side a. To locate vertex C, draw a circle of radius a with center B. Every point where this circle intersects the initial side of A is the vertex C of a triangle ABC that has sides a and c and angle A opposite side a.

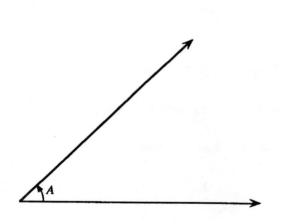

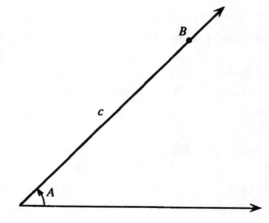

Figure 3.15   **Figure 3.16**

Locating vertex C is the critical step in this construction. The circle drawn to locate C might intersect the initial side of A in two points, one point, or not at all. Everything depends on how the length of side a compares with the distance between vertex B and the initial side of angle A. The expression for the distance between B and the initial side of A depends, in turn, on whether A is an acute or an obtuse angle. Let us continue investigating the case $A < 90°$ (shown in Figures 3.15 and 3.16) and return to the case $90° \leq A < 180°$ in a later paragraph.

When A is an acute angle, the distance h between vertex B, as constructed above, and the initial side of A is the *perpendicular* distance from the point B to the initial side of angle A. This is just the height h of the right triangle with base on the initial side of A, A as an acute angle and c as hypotenuse. Thus, $h = c \sin A$.

If the length of a is less than h (*i.e.*, $a < c \sin A$), then side a will not extend from vertex B to the initial side of A to enclose a triangle. There is no solution in this situation. Figure 3.17 illustrates this possibility.

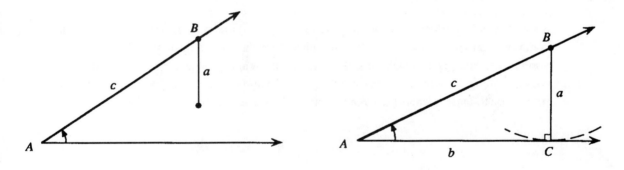

Figure 3.17 $a < c \sin A$ **Figure 3.18** $a = c \sin A$

If a is exactly equal to h (*i.e.*, $a = c \sin A$), the perpendicular line segment from vertex B to the initial side of A is side a and there is exactly one solution. In this case, C is a right angle and $\sin C = 1$. Figure 3.18 illustrates this possibility.

If the length of side a is greater than h but less than the length of side c, the circle of radius a centered at B intersects the initial side of A at two points. Thus, there are two points on the initial side of A that are distance a from B. Each one determines a vertex C for a triangle which has specified angle A and sides a and c, so there are two solutions. Figure 3.19 shows this construction.

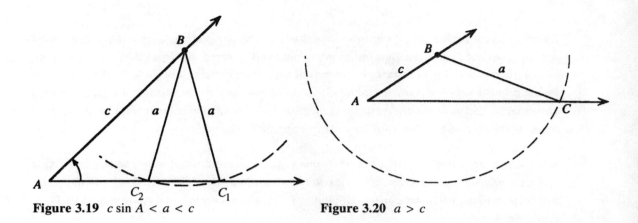

Figure 3.19 $c \sin A < a < c$ **Figure 3.20** $a > c$

Finally, consider the possibility that side a is longer than side c ($c \le a$). Because the initial side of A is a *ray*, not a line, now the circle of radius a centered at B intersects the initial side of angle A in only one point different from A. This point is again a vertex C for the only triangle that has specified angle A and sides a and c. Figure 3.20 shows this construction.

When $A \ge 90°$, the distance between the vertex B and the initial side of A is the straight line distance from vertex B to the initial point of the ray which forms the initial side of A. This point is the vertex A and this distance is the length of side c. Now, if the length of side a is no more than c, $(a \le c)$ side a will not be long enough to reach any point on the initial side of A (other than, possibly, A itself) to enclose a triangle. Hence, there is no solution when $A \ge 90°$ and $a \le c$. Figure 3.21 shows this case.

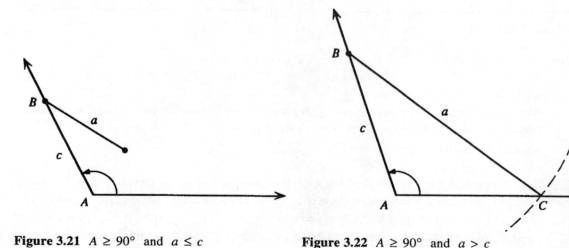

Figure 3.21 $A \ge 90°$ and $a \le c$ **Figure 3.22** $A \ge 90°$ and $a > c$

If side a is longer than side c ($c < a$), then the circle of radius a centered at B will intersect the initial side of A in only one point. This point C is the vertex of a triangle which has specified angle A and sides a and c. We have only one solution. Figure 3.22 shows this construction.

To solve an oblique triangle for which two sides and an opposite angle are given, we must calculate the measures of the angles and the lengths of the sides of the triangles we considered drawing in previous paragraphs. We need to know what happens computationally when a triangle has no solution, one solution or two solutions.

As before, suppose we are given sides a and c, and angle A opposite side a. We saw geometrically that the solution to the triangle hinges on the angle C. Think about how the computations of angle C might proceed. We have, by the Law of Sines,

$$\frac{a}{\sin A} = \frac{c}{\sin C}$$

so

$$\sin C = \frac{c \sin A}{a}.$$

The angle C is found by solving this inverse problem as in Objective 2.5.

Since $0° < A < 180°$, $\sin A > 0$. The numbers a and c are positive because they are lengths of line segments. Therefore, $\frac{c \sin A}{a} > 0$. However, $\frac{c \sin A}{a}$ can have any positive value. If $a < c \sin A$ so

$$\sin C = \frac{c \sin A}{a} > 1,$$

then, from Unit 2, there is no value for angle C and the triangle has no solution. This is the situation shown in Figure 3.17. If $a = c \sin A$ so

$$\sin C = \frac{c \sin A}{a} = 1,$$

then $\sin C = 1$, $C = 90°$ and the triangle has only one solution, a right triangle. This is the situation shown in Figure 3.18. If $0 < c \sin A < a$ so

$$0 < \sin C = \frac{c \sin A}{a} < 1,$$

then there are two possible values for the angle C. There is an acute angle C_1 and an obtuse angle C_2 (which has reference angle C_1). C_1 and C_2 may both be solutions as in Figure 3.19. C_1 may be the only solution as in Figures 3.20 and 3.22. Finally, neither C_1 nor C_2 may be solutions as in Figure 3.21. Fortunately, there is a simple procedure for deciding which of the angles C_1 and C_2 are solutions. Simply compute $A + C_1$ and $A + C_2$. If $A + C_1$ is strictly less than $180°$, C_1 is a solution. If $A + C_1$ is $180°$ or more, C_1 is not a solution. If $A + C_2$ is strictly less than $180°$, C_2 is a solution. If $A + C_2$ is $180°$ or more, C_2 is not a solution.

Summary

1. The problem of solving a triangle when two sides and an opposite angle are known is called the **ambiguous case** because it is not clear how many triangles having the given sides and angle there are. There may be no such triangles (no solution), exactly one such triangle (one solution) or two such triangles (two solutions).

2. By attempting to draw triangles which have two sides a and c of specified lengths and angle A of specified measure, we discover several results.

In the case that $0° < A < 90°$ we discover the following:

- If $a < c \sin A$, there is no such triangle (the problem has no solution).
- If $a = c \sin A$, there is one such triangle (the problem has one solution).
- If $c \sin A < a < c$, there are two such triangles (the problem has two solutions).
- If $c \le a$, there is one such triangle (the problem has one solution).

In the case that $90° \le A < 180°$ we discover the following:

- If $a \le c$, there is no such triangle (the problem has no solution).
- If $c < a$, there is one such triangle (the problem has one solution).

3. Given the lengths of two sides a and c and the measure of angle A opposite side a, calculate angle C of a triangle ABC which has the given sides and angle as follows.

First, substitute the given values for a, c and A into the equation

$$\frac{c}{\sin C} = \frac{a}{\sin A}$$

from the Law of Sines. Evaluate $\sin A$ and solve for $\sin C$ to find

$$\sin C = \frac{c \sin A}{a}.$$

The right side of this equation is a positive number calculated from c, A and a.

Second, solve for the angle(s) C that satisfy this equation and determine the corresponding triangle(s).

- If $\dfrac{c \sin A}{a} > 1$, then the equation has no solution and there is no triangle with specified sides and angle.

- If $\dfrac{c \sin A}{a} = 1$, then C is equal to $90°$ is the only solution to the equation and there is one triangle (a right triangle) with specified sides and angle.

- If $0 < \dfrac{c \sin A}{a} < 1$, then the equation has two solutions and there may be no, one or two triangles with specified sides and angle. An angle C which satisfies the equation is an angle of a triangle ABC with sides a and c and opposite angles A and C if and only if $A + C < 180°$.

Authors' Note

Ambiguous means susceptible of multiple interpretation, or doubtful or uncertain. Some synonyms are equivocal, obscure, recondite, abstruse, vague, cryptic and enigmatic. These variously may connote

- erudite obscurity of the scholar,
- lacking in clarity of meaning,
- puzzling terseness intended to discourage understanding,
- meaning hidden in difficult form, sometimes not worth digging out,
- deliberately unclear or misleading, suggesting a hedging to avoid exposure of one's position,
- lacking in definite form,
- great significance hidden in mysterious and challenging form and
- indicates the presence of two or more possible meanings, usually because of faulty expression.

Which of the definitions above do you feel applies best to the Law of Sines *ambiguous case*?

Examples

Example 1. Solve triangle *ABC*, where $a = 10$, $c = 15$ and $A = 43°$.

Solution 1.

Step 1. Sketch the triangle. Label the known and unknown quantities.

Step 2. Solve for angle *C* by using the Law of Sines.

$$\frac{c}{\sin C} = \frac{a}{\sin A}$$

$$\frac{15}{\sin C} = \frac{10}{\sin 43°}$$

$$\sin C = \frac{15}{10}\sin 43° = 1.0230.$$

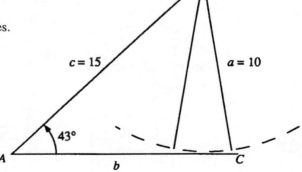

Figure 3.23

There is no angle *C* whose sine is greater than 1. There is **no** solution.

Example 2. Solve triangle ABC, where $a = 8.4$, $c = 10.5$ and $A = 53.13°$.

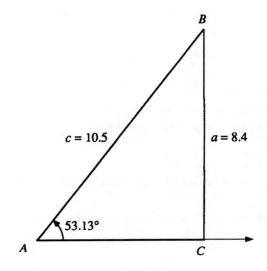

Solution 2.

Step 1. Sketch the triangle. Label the known and unknown quantities.

Step 2. Solve for angle C by using the Law of Sines.

$$\frac{c}{\sin C} = \frac{a}{\sin A}$$

$$\frac{10.5}{\sin C} = \frac{8.4}{\sin 53.13°}$$

$$\sin C = \frac{10.5}{8.4} \sin 53.13° = 1$$

$$C = 90° \text{ (see Objective 2.5)}$$

Figure 3.24

The triangle is a right triangle and there is only **one** solution.

Step 3. Solve for angle B by using $A + B + C = 180°$.

$$53.13° + B + 90° = 180°$$

$$B = 180° - 53.13° - 90°$$

$$B = 36.87°$$

Step 4. Solve for side b by the Pythagorean Theorem (or a trigonometric ratio).

$$a^2 + b^2 = c^2$$

$$(8.4)^2 + b^2 = (10.5)^2$$

$$b^2 = (10.5)^2 - (8.4)^2 = 39.69$$

$$b = \sqrt{39.69} = 6.3$$

The solution for the triangle is

$$a = 8.4, \qquad b = 6.3, \qquad c = 10.5,$$
$$A = 53.13°, \qquad B = 36.87°, \qquad C = 90°.$$

Example 3. Solve triangle ABC, where $a = 22$, $c = 27$ and $A = 50°$.

Solution 3.

Step 1. Sketch the triangle and label the known and unknown quantities.

Step 2. Solve for angle C by using the Law of Sines.

$$\frac{27}{\sin C} = \frac{22}{\sin 50°}$$

$$\sin C = 27 \cdot \frac{\sin 50°}{22} = 0.9401$$

$C_1 = 70.1°$, $C_2 = 109.9°$ (see Objective 2.5)

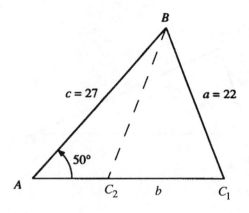

Figure 3.25

Step 3. Test the angles C_1 and C_2.

$$A + C_1 = 50° + 70.1° = 120.1° < 180°$$
$$A + C_2 = 50° + 109.9° = 159.9° < 180°$$

Since both sums are less than $180°$, $C_1 = 70.1°$ and $C_2 = 109.9°$ are solutions.

Step 4. Solve for angle B by using $A + B + C = 180°$. There are two possibilities.

$C_1 = 70.1°$ $\qquad\qquad\qquad\qquad\qquad$ $C_2 = 109.9°$

$50° + B_1 + 70.1° = 180°$ $\qquad\qquad\qquad$ $50° + B_2 + 109.9° = 180°$

$\qquad\quad B_1 = 59.9°$ $\qquad\qquad\qquad\qquad\qquad B_2 = 20.1°$

Step 5. Solve for side b by using the Law of Sines. There are two possibilities.

$C_1 = 70.1°$ $\qquad\qquad\qquad\qquad\qquad\qquad$ $C_2 = 109.9°$

$$\frac{b_1}{\sin 59.9°} = \frac{22}{\sin 50°} \qquad\qquad\qquad \frac{b_2}{\sin 20.1°} = \frac{22}{\sin 50°}$$

$$b_1 = \sin 59.9° \cdot \frac{22}{\sin 50°} = 24.85 \qquad b_2 = \sin 20.1° \cdot \frac{22}{\sin 50°} = 9.87$$

The two solutions for the triangle are

$$\left.\begin{matrix} a = 22 & b_1 = 24.85 & c = 27 \\ A = 50° & B_1 = 59.9° & C_1 = 70.1° \end{matrix}\right\} \text{ and } \begin{cases} a = 22 & b_2 = 9.87 & c = 27 \\ A = 50° & B_2 = 20.1° & C_2 = 109.9°. \end{cases}$$

Example 4. Solve triangle ABC, where $a = 13$, $c = 6$ and $A = 70°$.

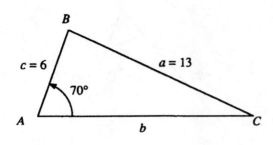

Figure 3.26

Solution 4.

Step 1. Sketch the triangle and label the known and unknown quantities.

Step 2. Solve for angle C by the use of the Law of Sines.

$$\frac{6}{\sin C} = \frac{13}{\sin 70°}$$

$$\sin C = 6 \cdot \frac{\sin 70°}{13} = 0.4337$$

$$C_1 = 25.7°, \quad C_2 = 154.3°$$

Step 3. Test the angles C_1 and C_2.

$$A + C_1 = 70° + 25.7° = 95.7° < 180°$$
$$A + C_2 = 70° + 154.3° = 224.3° \geq 180°$$

Since $A + C_1 < 180°$, C_1 is a solution.
Since $A + C_2 \geq 180°$, C_2 is not a solution.

$C_1 = 25.7°$ is the only solution.

Step 4. Solve for angle B by using $A + B + C = 180°$.
$$70° + B + 25.7° = 180°$$
$$B = 84.3°$$

Step 5. Solve for side b by using the Law of Sines.

$$\frac{b}{\sin 84.3°} = \frac{13}{\sin 70°}$$

$$b = \sin 84.3° \cdot \frac{13}{\sin 70°} = 13.77$$

The solution for the triangle is

$$a = 13, \qquad b = 13.77, \qquad c = 6,$$
$$A = 70°, \qquad B = 84.3°, \qquad C = 25.7°.$$

Practice Problems

1. Solve the triangle ABC, where $a = 221$, $c = 543$ and $A = 23°$.

2. Solve the triangle ABC, where $a = 89.1$, $c = 100.0$ and $A = 63°$.

3. Solve the triangle ABC, where $A = 39°$, $a = 40$ and $c = 47$.

4. Solve the triangle ABC, where $a = 5.8$, $c = 6.2$ and $A = 63°$.

5. Solve the triangle ABC, where $a = 49.8$, $c = 57.5$ and $A = 136°$.

6. Solve the triangle ABC, where $a = 9.2$, $c = 7.6$ and $A = 98.6°$.

7. Solve the triangle ABC, where $a = 17.2$, $c = 21.6$ and $A = 121.7°$.

8. Solve the triangle ABC, where $B = 72°$, $b = 8.1$ and $c = 8.3$.

9. Solve the triangle PQR, where $R = 14.2°$, $q = 17.3$ and $r = 15.7$.

10. Solve the triangle GEF, where $E = 136°$, $e = 111$ and $f = 57.5$.

11. Find angle B in the triangle ABC, where $a = 1$, $b = 8$ and $A = 52°$.

12. Find side t in the triangle RST, where $r = 17$, $s = 8.5$ and $S = 30°$.

13. Solve the triangle DRT, where $r = 311.9$, $t = 205.3$ and $R = 55.5°$.

14. Solve the triangle BAD, where $A = 105.7°$, $a = 32.9$ and $b = 46.8$.

15. Find the perimeter of the triangle WAK, where $a = 31.2$, $k = 49$ and $A = 32.7°$.

16. Find angle C in triangle ABC, where $a = 317$, $b = 465$ and $A = 36°$.

17. Solve the triangle BCD, where $c = 96$, $b = 70$ and $C = 147°$.

18. Find the perimeter of triangle SUN, which has $U = 9.6°$, $s = 21.6$ and $u = 18.5$.

19. In oblique triangle ABC, $a = 3.6$, $b = 8.1$ and $A = 58°$. Which one of the following might be angle B to the nearest tenth of a degree? Choose all correct responses.

(A) $22.1°$ (B) $37.7°$ (C) $65.2°$ (D) $87.6°$ (E) none of these

20. In oblique triangle ABC, $a = 0.7$, $c = 2.4$ and $C = 98°$. Which one of the following might be angle A to the nearest tenth of a degree? Choose all correct responses.

(A) $3.4°$ (B) $28.8°$ (C) $73.2°$ (D) $163.2°$ (E) none of these

---

UNIT 3
Sample Examination 1

1. What is the (approximate) perimeter of right triangle ACB, where $a = 51.13$, $B = 5°$ and $C = 90°$?

(A) 106.90   (B) 106.93   (C) 106.96   (D) 107.01   (E) 107.04

2. In right triangle ACB, $a = 0.52$, $c = 0.59$ and $C = 90°$. What is the (approximate) value of $B - A$?

(A) −34.7°   (B) −34.4°   (C) −34.2°   (D) −33.9°   (E) −33.6°

3. What is the (approximate) length of side q of right triangle PQR, where $p = 9.13$, $r = 3.91$ and $Q = 90°$?

(A) 9.59   (B) 9.68   (C) 9.75   (D) 9.84   (E) 9.93

4. To the nearest unit, what is the length of side a of (possibly oblique) triangle ABC, where $A = 89°$, $B = 17°$ and $c = 800$? There is at least one correct response. Choose all correct responses.

(A) 812   (B) 817   (C) 821   (D) 827   (E) 832

5. In (possibly oblique) triangle ABC, $a = 316$, $c = 592$ and $C = 67°$. Which one of the following might be angle A to the nearest tenth of a degree? There is at least one correct response. Choose all correct responses.

(A) 29.4°   (B) 96.4°   (C) 148.5°   (D) 150.6°   (E) none of these

UNIT 3
Sample Examination 2

1. What is the (approximate) perimeter of right triangle ABC, where $c = 0.42$, $A = 82°$ and $C = 90°$?

(A) 0.84   **(B)** 0.89   **(C)** 0.93   **(D)** 0.97   **(E)** 1.01

2. In right triangle ACB, $a = 12.39$, $b = 46.76$ and $C = 90°$. What is the (approximate) value of $A - B$?

(A) $-61.2°$ **(B)** $-60.9°$ **(C)** $-60.3°$ **(D)** $-60.0°$ **(E)** $-59.4°$

3. What is the (approximate) length of side h of right triangle HJP, where $j = 9.03$, $p = 3.61$ and $J = 90°$?

(A) 8.28   **(B)** 8.36   **(C)** 8.38   **(D)** 8.46   **(E)** 8.48

4. To the nearest tenth, what is the length of side b of (possible oblique) triangle ABC, where $B = 54°$, $C = 94°$ and $a = 24.6$? There is at least one correct response. Choose all correct responses.

(A) 37.6   **(B)** 38.9   **(C)** 40.2   **(D)** 41.5   **(E)** none of these

5. In (possibly oblique) triangle ABC, $b = 731.6$, $c = 683.9$ and $C = 118°$. Which of the following might be angle B to the nearest tenth of a degree? There is at least one correct response. Choose all correct responses.

(A) 59.1°   **(B)** 60.6°   **(C)** 62.1°   **(D)** 70.8°   **(E)** none of these

UNIT 4

THE LAW OF COSINES AND APPLIED PROBLEMS

Introduction

In this Unit we will complete the study of the solution of oblique triangles by introducing the Law of Cosines. We will solve arbitrary triangles where it is necessary to determine what method to use: trigonometric functions for right triangles, the Law of Sines for Case I or Case II oblique triangles, or the Law of Cosines for Case III or Case IV oblique triangles. We also will solve a variety of applied problems whose geometry is that of a triangle.

100

UNIT 4
THE LAW OF COSINES AND APPLIED PROBLEMS

Objective 4.1

(a) Given two sides and the included angle of a (possibly oblique) triangle, solve the triangle using the Law of Cosines.

(b) Given three sides of a (possibly oblique) triangle, solve the triangle using the Law of Cosines.

Objective 4.2

Given three parts of a triangle (including at least one side),

(a) determine what method (right triangle, Law of Sines or Law of Cosines) should be used to solve the triangle, and

(b) solve the triangle.

Objective 4.3

Given an applied problem whose geometry is that of a triangle, solve the problem.

THE LAW OF COSINES AND APPLIED PROBLEMS

Objective 4.1

(a) Given two sides and the included angle of a (possibly oblique) triangle, solve the triangle using the Law of Cosines.

(b) Given three sides of a (possibly oblique) triangle, solve the triangle using the Law of Cosines.

Discussion

The Law of Sines relates two angles of a triangle with the sides opposite those angles. In contrast, **the Law of Cosines relates one angle with the three sides of a triangle.** In terms of the standard labeling of a triangle ABC, the Law of Cosines says that

$$a^2 = b^2 + c^2 - 2bc \cos A .$$

This relationship is needed to solve triangles for which two sides and the included angle (Case III) or three sides (Case IV) are given.

To see why the Law of Cosines is true, consider a general oblique triangle ABC as shown in Figure 4.1. Although angle A is drawn as an obtuse angle, the argument we give is easily modified in the event angle A is acute.

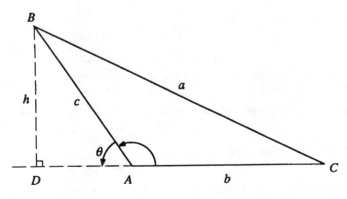

Figure 4.1

From vertex B construct a perpendicular BD to the line containing AC. Let h denote the length BD. By the Pythagorean Theorem, in right triangle BDC we have

$$a^2 = h^2 + (DC)^2,$$

where the symbol DC denotes the length of the line segment joining points D and C. Since $DC = DA + b$, we can substitute $DA + b$ for DC and obtain

(1)
$$a^2 = h^2 + (DA + b)^2 = h^2 + (DA)^2 + 2b(DA) + b^2.$$

By the Pythagorean Theorem, in right triangle BDA

(2)
$$h^2 + (DA)^2 = c^2.$$

Also, the angle θ is the reference angle for angle A, so $\cos \theta = -\cos A$. Consequently,

(3)
$$\frac{DA}{c} = \cos \theta = -\cos A \quad \text{and} \quad DA = -c \cos A.$$

Finally, substituting equations (2) and (3) into equation (1) gives

$$a^2 = c^2 + 2b(-c \cos A) + b^2$$

or

(4)
$$a^2 = b^2 + c^2 - 2bc \cos A.$$

Equation (4) is one form of the Law of Cosines. It expresses side a in terms of the opposite angle A and the sides b and c. Similar equations express sides b and c in terms of their opposite angles and the remaining sides. We summarize these conclusions as a theorem.

THEOREM   **Law of Cosines:**
In any triangle ABC,

$$a^2 = b^2 + c^2 - 2bc \cos A,$$
$$b^2 = a^2 + c^2 - 2ac \cos B,$$
$$c^2 = a^2 + b^2 - 2ab \cos C,$$

where a, b and c are the lengths of the sides opposite the angles A, B and C.

When $\triangle ACB$ is a right triangle with right angle C, $\cos C = 0$, so the equation

$$c^2 = a^2 + b^2 - 2ab \cos C,$$

from the Law of Cosines, becomes

$$c^2 = a^2 + b^2.$$

This is just the Pythagorean relation. Thus, the Pythagorean Theorem is a special case of the Law of Cosines.

The Law of Cosines can be used directly to find the third side of a triangle for which two sides and the included angle are known (Case III). Just write the Law of Cosines in the form which involves the known angle and sides, substitute the known values for the two sides and the included angle, and compute the third side.

Now turn to the problem of finding a triangle whose three sides have given lengths a, b and c (Case IV). We know that there may or may not be such a triangle. In any triangle the sum of the lengths of any two of the sides must exceed the length of the third side. Thus, if any one of the numbers a, b and c is greater than or equal to the sum of the other two, there is no triangle with sides of lengths a, b and c. On the other hand, if each of the numbers a, b and c is less than the sum of the other two, we can draw a triangle with sides of these lengths with a ruler and compass. (Draw a line segment AB of length c. Draw a circle of radius b centered at endpoint A and a circle of radius a centered at endpoint B. Since $a + b > c$, these circles intersect. Use either intersection point as vertex C of a triangle ABC. This triangle has sides of lengths a, b and c.)

The Law of Cosines provides the tool we need to calculate the measures of the angles of this triangle. The equation

$$a^2 = b^2 + c^2 - 2bc \cos A$$

tells us how angle A is related to the lengths of the three sides. (And, of course, the other forms of the Law of Cosines tell us how angles B and C are related to a, b and c.) By solving for $\cos A$ we find

(5)
$$\cos A = \frac{b^2 + c^2 - a^2}{2bc}.$$

We can find A from this equation by substituting the values for a, b and c, and then solving the resulting inverse problem. Because this problem involves the cosine and because we need only the angle A between $0°$ and $180°$ (rather than the two angles we sought in Objective 2.5), this inverse problem is especially easy to solve with the calculator. With a basic scientific calculator, set the calculator to the desired mode (normally degrees), enter the value of the cosine as calculated from equation (5), and press [INV] and [COS]. (You may be able to use the calculator's memory to avoid re-entering numbers you have calculated. Consult the operator's manual.) With an advanced scientific or graphics calculator, set the calculator to the desired mode (normally degrees), press [2nd] and [COS⁻¹], enter the value of the cosine as calculated from equation (5), and press the [ENTER] key. (You may be able to use the calculator's ANS function to avoid re-entering numbers you have calculated. Consult the operator's manual.) Either calculator will produce the desired angle A between $0°$ and $180°$ (no need for reference angles this time).

Find a second angle of the triangle by repeating this calculation with the other forms of the Law of Cosines, and find the third angle by using the fact that the measures of the angles must sum to $180°$.

The Law of Sines can also be used to find the remaining angles, but not without risks. Finding an angle from two sides and an opposite angle, as required in this situation, is the ambiguous case. The calculations may produce two angles, one acute and one obtuse. The obtuse angle is easily overlooked, but when triangle ABC is obtuse, this angle may be the solution to the triangle. The Law of Sines is especially efficient for calculator computations, but must be used critically.

In quadrant I $0 < \cos \theta < 1$ and in quadrant II $-1 < \cos \theta < 0$. Consequently, when $0 < \cos \theta = \dfrac{b^2 + c^2 - a^2}{2bc} < 1$, θ is an acute angle, when $\cos \theta = \dfrac{b^2 + c^2 - a^2}{2bc} = 0$, θ is a right angle, and when $-1 < \cos \theta = \dfrac{b^2 + c^2 - a^2}{2bc} < 0$, θ is an obtuse angle. When there is no triangle with sides a, b and c (i.e., when one of the numbers a, b and c is greater than or equal to the sum of the other two),

$$\left| \frac{a^2 + b^2 - c^2}{2ab} \right| \geq 1, \quad \left| \frac{b^2 + c^2 - a^2}{2bc} \right| \geq 1, \quad \text{and} \quad \left| \frac{a^2 + c^2 - b^2}{2ac} \right| \geq 1.$$

Every inverse problem that arises from trying to find such a triangle tells us there is none.

Summary

1. To solve a triangle for which two sides and the included angle are known, write the Law of Cosines in the form which involves the known angle and sides, substitute the known values and compute the third side.

2. There is a triangle whose sides have lengths $a > 0$, $b > 0$ and $c > 0$ if and only if each one of the numbers a, b and c is less than the sum of the other two.

3. Use the Law of Cosines to determine whether there is a triangle ABC with sides of given lengths $a > 0$, $b > 0$ and $c > 0$ and, if there is such a triangle, to find its angles. Solve the equation for the Law of Cosines that involves angle A for $\cos A$ and obtain

$$\cos A = \frac{b^2 + c^2 - a^2}{2bc}.$$

There is a triangle with sides of the given lengths if and only if $\left| \dfrac{b^2 + c^2 - a^2}{2bc} \right| < 1$. When $\triangle ABC$ exists, use a scientific calculator to solve for A by applying the inverse cosine to the numerical value of $\dfrac{b^2 + c^2 - a^2}{2bc}$ (complete with sign). Compute the remaining angles in this way from the equations for the Law of Cosines involving the other angles or by using methods studied previously. The Law of Sines is especially efficient for calculating these angles, but must be used critically.

Examples The following Examples illustrate the application of the Law of Cosines.

Example 1. (Case III: two sides and the included angle are given.) Solve triangle ABC, where $a = 7$, $b = 9$ and $C = 47°$.

Solution 1.

Step 1. Sketch the triangle and label the known and unknown quantities.

Step 2. Solve for side c by using the Law of Cosines.

$$c^2 = a^2 + b^2 - 2ab \cos C$$
$$c^2 = 7^2 + 9^2 - 2(7)(9) \cos 47°$$
$$c = \sqrt{44.0682} = 6.64$$

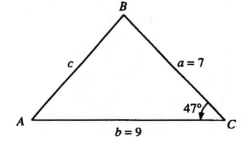

Figure 4.2

Step 3. Solve for angle A by using the Law of Cosines.

$$a^2 = b^2 + c^2 - 2bc \cos A$$

$$7^2 = 9^2 + 6.64^2 - 2(9)(6.64) \cos A$$

$$\cos A = \frac{9^2 + 6.64^2 - 7^2}{2(9)(6.64)} = 0.6366$$

$$A = 50.5°$$

Step 4. Solve for angle B by using $A + B + C = 180°$.

$$50.5° + B + 47° = 180°$$

$$B = 180° - 50.5° - 47° = 82.5°$$

The complete solution of the triangle is

$$a = 7, \qquad b = 9, \qquad c = 6.64,$$
$$A = 50.5°, \qquad B = 82.5°, \qquad C = 47°.$$

Example 2. (Case IV: three sides are given.) Solve triangle ABC, where $a = 6$, $b = 3$ and $c = 5$.

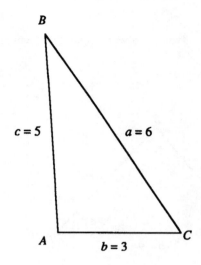

Figure 4.3

Solution 2.

Step 1. Sketch the triangle and label the known and unknown quantities.

Step 2. Solve for angle A by using the Law of Cosines.

$$a^2 = b^2 + c^2 - 2bc \cos A$$
$$6^2 = 3^2 + 5^2 - 2(3)(5) \cos A$$
$$\cos A = \frac{3^2 + 5^2 - 6^2}{2(3)(5)} = -0.0667$$
$$A = 93.8°$$

Step 3. Solve for angle B by using the Law of Cosines.

$$b^2 = a^2 + c^2 - 2ac \cos B$$
$$3^2 = 6^2 + 5^2 - 2(6)(5) \cos B$$
$$\cos B = \frac{6^2 + 5^2 - 3^2}{2(6)(5)} = 0.8667$$
$$B = 29.9°$$

Step 4. Solve for angle C by using $A + B + C = 180°$.
$$93.8° + 29.9° + C = 180°$$
$$C = 180° - 93.8° - 29.9° = 56.3°$$

The Law of Cosines could also be used to solve for angle C.

The complete solution of the triangle is

$$a = 6, \qquad b = 3, \qquad c = 5,$$
$$A = 93.8°, \qquad B = 29.9°, \qquad C = 56.3°.$$

Example 3. (Case IV: three sides are given.) Solve triangle ABC, where $a = 1$, $b = 4$ and $c = 2$.

Solution 3.

Step 1. Sketch the triangle and label the known and unknown quantities. Attempts to accurately draw the triangle show that since

$$a + c = 3 < 4 = b,$$

the three segments do not form a triangle. Consequently, there is **no solution.**

Attempts to find A, B and C by the Law of Cosines will give values for $\cos A$, $\cos B$ and $\cos C$ which are not between -1 and 1. For instance,

$$\cos A = \frac{b^2 + c^2 - a^2}{2bc} = \frac{16 + 4 - 1}{16} = \frac{19}{16} > 1.$$

This, too, indicates there is **no solution.**

Practice Problems

1. Find side a of the triangle in Figure 4.4.

2. Find angle A of the triangle in Figure 4.5.

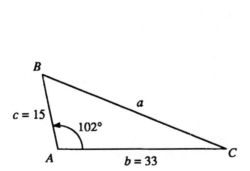

Figure 4.4

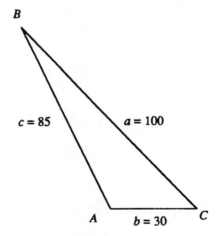

Figure 4.5

3. Solve the triangle ABC, where $a = 2$, $b = 3$ and $c = 7$.

4. Solve the triangle ABC, where $a = 0.8$, $c = 1.7$ and $B = 28.07°$.

5. Find side a in the triangle CAT, where $c = 325$, $t = 325$ and $A = 70°$.

6. Find angle R in the triangle PQR, where $p = 53$, $q = 101$ and $r = 95$.

7. Solve the triangle PAS, where $p = 7$, $a = 24$ and $s = 25$.

8. Solve the triangle XYZ, where $x = 1.5$, $y = 3.6$ and $z = 3.9$.

9. Find angle A in triangle ABC, where $a = 39$, $c = 12.9$ and $B = 57°$.

10. Solve the triangle ABC, where $a = 7$, $b = 7$ and $c = 7$.

11. Two sides of a triangle have lengths 51 and 29. The included angle measures 121°. Solve the triangle.

12. Two sides of a triangle have lengths 236 and 179. The included angle measures 42.3°. Solve the triangle.

13. Three sides of a triangle have lengths 0.8, 2.1 and 1.7. Solve the triangle.

14. Three sides of a triangle have lengths 21.9, 13.4 and 35.6. Solve the triangle.

15. Two sides of a triangle have lengths 1.4 and 5.0. The included angle measures 73.74°. Solve the triangle.

Objective 4.2

Given three parts of a triangle (including at least one side),
(a) determine what method (right triangle, Law of Sines, or Law of Cosines) should be used to solve the triangle, and
(b) solve the triangle.

Discussion

To solve an arbitrary triangle for which three parts, including one side, are given, first organize the given information by sketching the triangle and labeling the known and unknown parts. Use your drawing to evaluate the information and decide which method to use to solve the triangle. If a 90° angle is given or if two angles which sum to 90° are given, the triangle is a right triangle and should be solved by using the sum of angles relation, the Pythagorean relation and the ratios which define the trigonometric functions. If one side and two angles (which do not sum to 90°) are known (Case I), the triangle is completely determined and the Law of Sines will yield the unknown sides. If two sides and an angle opposite one of them are known (Case II), the triangle may have no solution, one solution, or two solutions. Use the Law of Sines (ambiguous case) to analyze the situation and find all possible solutions. Two sides and the included angle (Case III) also determine a triangle completely, and the Law of Cosines can be used to solve for the third side. Various methods can then be used to find the missing angles. Three sides (Case IV) may or may not determine a triangle. Use the Law of Cosines to decide whether there is such a triangle and, if there is, to find its angles.

Authors' Note

One of the most significant mathematics textbook ever written is <u>The Elements</u> (*Stoichia*) by Euclid (circa. 300 B.C.). In the second book of <u>The Elements</u> the reader can find the formulations, for both the obtuse and the acute angles, of what later became known as the Law of Cosines for plane triangles. Considering the historical impact of this book, not much is known of Euclid's life. There is a tale told of him in his position as a teacher in Alexandria, however, in which his response to a student's inquiry, questioning whether there was a use for the study of geometry, was to instruct that the student be given money, "since he must needs make gain of what he learns." Don't try this!

Examples

Example 1. Determine what method should be used to solve triangle ABC, where $b = 72$, $c = 48$ and $B = 90°$. Then solve the triangle.

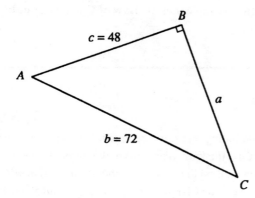

c = 48

B

A

a

b = 72

C

Figure 4.6

Solution 1.

Step 1. Sketch the triangle and label the known and unknown quantities as in Figure 4.6.

Step 2. Determine the method of solution. Since B is a right angle, this is a right triangle. It is best solved by using trigonometric ratios as in Objective 3.1. Notice that in this triangle the right angle is labeled B (instead of C) so the hypotenuse is labeled b (instead of c). Be careful to use the trigonometric ratios cor-rectly.

Step 3. Solve for side a by using the Pythagorean relation.

$$a^2 + c^2 = b^2$$
$$a^2 + (48)^2 = (72)^2$$
$$a^2 = (72)^2 - (48)^2 = 2880$$
$$a = 53.67$$

Step 4. Solve for angle A by using a trigonometric ratio.

$$\cos A = \frac{\text{side adjacent to } A}{\text{hypotenuse}} = \frac{48}{72} = 0.6667$$
$$A = 48.2°$$

Step 5. Solve for angle C by using $A + C = 90°$.

$$48.2° + C = 90°$$
$$C = 90° - 48.2° = 41.8°$$

Alternately, angle C could be found by using trigonometric ratios.

The complete solution of the triangle is

$$a = 53.67, \qquad b = 72, \qquad c = 48,$$
$$A = 48.2°, \qquad B = 90°, \qquad C = 41.8°.$$

Example 2. Determine what method should be used to solve triangle ABC, where $a = 13$, $c = 4$ and $A = 122°$. Then solve the triangle.

Solution 2.

Step 1. Sketch the triangle and label the known and unknown quantities as in Figure 4.7.

Step 2. Determine the method of solution. Since two sides and an opposite angle are given (Case II), this triangle should be solved using the Law of Sines as in Objective 3.3. It is evident from the drawing that, since A is an obtuse angle and $a > c$, C must be an acute angle and this triangle has exactly one solution.

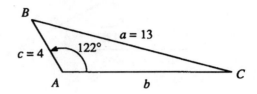

Figure 4.7

Step 3. Solve for angle C by using the Law of Sines.

$$\frac{4}{\sin C} = \frac{13}{\sin 122°}$$

$$\sin C = \frac{4 \sin 122°}{13} = 0.2609$$

$$C = 15.1°$$

This inverse problem has two solutions between $0°$ and $180°$. One solution is an acute angle, the other obtuse. Since we know from Step 2 that C is an acute angle, there is no need to find the second solution.

Step 4. Solve for angle B by using $A + B + C = 180°$.

$$122° + B + 15.1° = 180°$$

$$B = 180° - 122° - 15.1° = 42.9°$$

Step 5. Solve for side b by using the Law of Sines.

$$\frac{b}{\sin 42.9°} = \frac{13}{\sin 122°}$$

$$b = \frac{13 \sin 42.9°}{\sin 122°} = 10.43$$

Alternately, side b could be found by using the Law of Cosines, but that would require more complicated calculations. Try it!

The complete solution of the triangle is

$$a = 13, \qquad b = 10.43, \qquad c = 4,$$
$$A = 122°, \qquad B = 42.9°, \qquad C = 15.1°.$$

Example 3. Solve triangle ABC, where $b = 23$, $B = 125°$ and $C = 30°$.

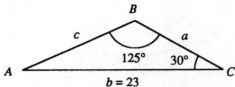

Solution 3.

Step 1. Sketch the triangle and label the known and unknown quantities.

Step 2. Determine the method of solution. One side and two angles are given (Case I). Use the Law of Sines to solve the triangle as in Objective 3.2.

Figure 4.8

Step 3. Solve for angle A by using $A + B + C = 180°$.

$$A + 125° + 30° = 180°$$

$$A = 180° - 125° - 30° = 25°$$

Step 4. Solve for side a by using the Law of Sines.

$$\frac{a}{\sin A} = \frac{b}{\sin B}$$

$$\frac{a}{\sin 25°} = \frac{23}{\sin 125°}$$

$$a = \sin 25° \cdot \frac{23}{\sin 125°} = 11.87$$

Step 5. Solve for side c by using the Law of Sines.

$$\frac{c}{\sin C} = \frac{b}{\sin B}$$

$$\frac{c}{\sin 30°} = \frac{23}{\sin 125°}$$

$$c = \sin 30° \cdot \frac{23}{\sin 125°} = 14.04$$

The complete solution of the triangle is

$$a = 11.87, \qquad b = 23, \qquad c = 14.04,$$
$$A = 25°, \qquad B = 125°, \qquad C = 30°.$$

Example 4. Solve triangle ABC, where $a = 61$, $c = 39$ and $B = 40°$.

Solution 4.

Step 1. Sketch the triangle and label the known and unknown quantities.

Step 2. Determine the method of solution. Two sides and an included angle are given (Case III). Use the Law of Cosines to solve the triangle as in Objective 4.1.

Step 3. Solve for side b by using the Law of Cosines.
$$b^2 = a^2 + c^2 - 2ac \cos B$$
$$b^2 = 61^2 + 39^2 - 2(61)(39) \cos 40° = 1597.16$$
$$b = 39.96$$

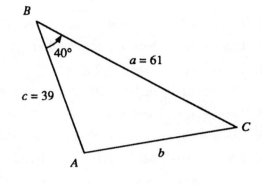

Figure 4.9

Step 4. Solve for angle A by using the Law of Cosines.
$$a^2 = b^2 + c^2 - 2bc \cos A$$
$$61^2 = 39.96^2 + 39^2 - 2(39.96)(39) \cos A$$
$$\cos A = \frac{39.96^2 + 39^2 - 61^2}{2(39.96)(39)} = -0.1935$$
$$A = 101.1°$$

Step 5. Solve for angle C by using $A + B + C = 180°$.
$$101.1° + 40° + C = 180°$$
$$C = 180° - 101.1° - 40° = 38.9°$$

The complete solution of the triangle is

$$a = 61, \qquad b = 39.96, \qquad c = 39,$$
$$A = 101.1°, \qquad B = 40°, \qquad C = 38.9°.$$

Practice Problems

In **Problems 1 - 18,** determine what method should be used to solve the given triangles, and then solve the triangles.

1. Triangle ABC, where $b = 36$, $c = 12$ and $B = 90°$.

2. Triangle ABC, where $a = 15$, $c = 6$ and $A = 127°$.

3. Triangle ABC, where $a = 9.5$, $A = 90°$ and $B = 62°$.

4. Triangle ABC, where $a = 28$, $b = 21$ and $c = 35$.

5. Triangle ABC, where $A = 31°$, $C = 90°$ and $a = 23$.

6. Triangle ABC, where $A = 19°$, $C = 46°$ and $c = 31$.

7. Triangle ABC, where $a = 17$, $b = 30$ and $C = 90°$.

8. Triangle ABC, where $a = 10$, $b = 11$ and $C = 130°$.

9. Triangle RST, where $r = 18$, $t = 14$ and $T = 35°$.

10. Triangle PQR, where $p = 300$, $q = 271$ and $r = 415$.

11. Triangle BCD, where $b = 7$, $d = 9$ and $B = 70°$.

12. Triangle XYZ, where $X = 58°$, $y = 7$ and $z = 15$.

13. Triangle RAT, where $T = 90°$, $A = 22°$ and $t = 10.5$.

14. Triangle MAD, where $m = 27$, $a = 16$ and $d = 40$.

15. Triangle PDQ, where $P = 53°$, $D = 88°$ and $q = 9.7$.

16. Triangle BAC, where $A = 90°$, $B = 77°$ and $c = 41$.

17. Triangle TWA, where $T = 13°$, $w = 47$ and $t = 52$.

18. Triangle MNP, where $M = 29°$, $n = 8$ and $p = 31$.

Objective 4.3

Given an applied problem whose geometry is that of a triangle, solve the problem.

Discussion

Applied problems, or *story problems*, are among the most interesting in mathematics because they suggest how mathematics is used to solve problems that arise in practice. At the same time, applied problems are especially challenging because each one has its own individual characteristics and no set procedure can be used to solve every problem. Experience has shown, however, that the following steps provide a systematic, and usually successful, approach to solving applied problems in trigonometry.

> **Step 1. Read** the problem. Look up unfamiliar terms. Then, read the problem again. Finally, to be sure you understand the problem, restate it accurately in your own words and check your restatement against the original.
>
> **Step 2. Draw** a sketch for the problem in order to clearly visualize the relationships among the pieces of information that are known and desired. Check your drawing against the original statement of the problem.
>
> **Step 3. Identify** and **label** the known and unknown quantities in your drawing. Identify the quantity (or quantities) to be found.
>
> **Step 4. Write** an equation (or equations) relating the known quantities with the quantity (or quantities) to be found.
>
> **Step 5. Solve** the equation(s) for the unknown quantity (or quantities).

The problems in this Objective will always involve a triangle. In these problems the equations(s) will come from trigonometric ratios for right triangles, the Pythagorean relation, the Law of Sines and/or the Law of Cosines. It is usually easier to obtain the equations called for in Step 4 if the triangle drawn in Step 2 is labeled in the standard way with angles denoted by capital letters and sides opposite denoted by the same letter in lower case.

Applications in this book include problems from surveying and navigation. We will need some terminology from these fields. If an observer views an object which is above him/her, then the positive acute angle between the line of sight and the horizontal plane is called the **angle of elevation.** For an object below an observer, the positive acute angle between the line of sight and the horizontal plane is called the **angle of depression.** These are illustrated in Figures 4.10 (a) and 4.10 (b).

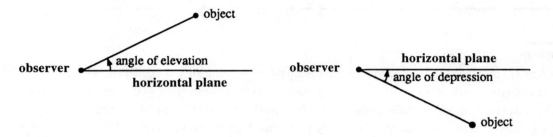

Figure 4.10 (a) angle of elevation   **Figure 4.10 (b)** angle of depression

Bearing is often used to specify a direction from a certain point. The bearing of a direction from a point O is given in terms of the acute angle between the north-south line through O and the ray from O in that direction. To specify a direction by the bearing, first, write the letter N or S (to indicate the reference direction north or south). Second, write the angle of deviation from north or south. Finally, write the letter E or W (to indicate whether the deviation is to the east or the west). For example, the bearing illustrated by Figure 4.11 (a) below, is N30°W. Figure 4.11 (b) through Figure 4.11 (d) illustrate several other bearings.

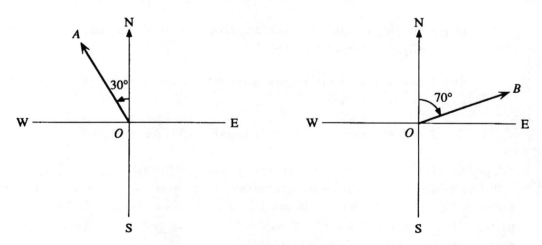

Figure 4.11 (a) Bearing of A from O is N30°W   **Figure 4.11 (b)** Bearing of B from O is N70°E

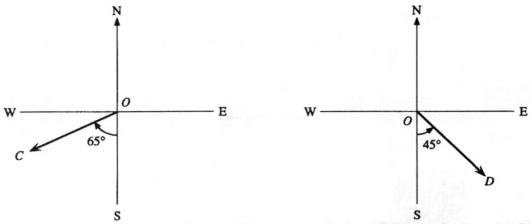

Figure 4.11 (c) Bearing of C from O is S65°W   **Figure 4.11 (d)** Bearing of D from O is S45°E

When the bearing from P to Q is known, it is easy to obtain the bearing from Q to P. Merely interchange the letters N and S and interchange the letters E and W. Thus, in Figures 4.11, the bearing of O from A is S30°E, the bearing of O from B is S70°W, the bearing of O from C is N65°E and the bearing of O from D is N45°W.

Examples

Example 1. The Washington Monument casts a shadow of length 793 feet when the angle of elevation of the sun is 35°. How tall is the monument? (For purposes of this problem, assume, incorrectly, that the ground around the Washington monument is level.)

Solution 1.
Step 1. Read the problem carefully.
Step 2. Draw a sketch for the problem.
Step 3. Identify and **label** the known and unknown quantities. Label the length of the shadow 793 ft. and the angle of elevation 35°. We want to find the height of the monument which we label x.
Step 4. Write an equation relating the known quantities with the height x of the monument. This is a right triangle, so this equation comes from trigonometric ratios.

$$\tan 35° = \frac{x}{793}$$

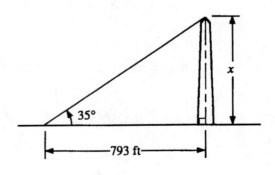

Figure 4.12

Step 5. Solve for x.

$$x = 793 \tan 35° = 793(0.7002) = 555.26 \text{ feet}$$

Example 2. To find the length of a lake, a point is chosen 2547 yards from one end of the lake and 3264 yards from the other end. At this point the angle subtended by the lake is 82.9°. How long is the lake?

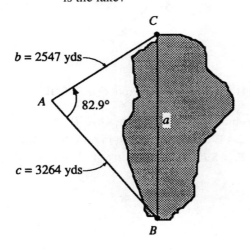

b = 2547 yds

A 82.9°

c = 3264 yds

C

a

B

Solution 2.
Step 1. Read the problem carefully.
Step 2. Draw a sketch for the problem.
Step 3. Identify and **label** the known and unknown quantities. Label the vertices of the triangle A, B and C so that angle A is 82.9°. Then, side b is 2547 and side c is 3264. We want to find the length of the lake which is side a.

Figure 4.13

Step 4. Write an equation relating the known quantities (two sides and the included angle) with the quantity to be found (the third side). The Law of Cosines applies and we have

$$a^2 = b^2 + c^2 - 2bc \cos A$$

$$= (2547)^2 + (3264)^2 - 2(2547)(3264) \cos 82.9°.$$

Step 5. Solve for a.

$$a^2 = 6,487,209 + 10,653,696 - 2,055,099 = 15,085,806$$

$$a = \sqrt{15,085,806} = 3884 \text{ yards}$$

Example 3. Two men are standing 500 feet apart on level ground and observe a hot air balloon between them. The respective angles of elevation of the balloon are measured by the men from ground level and found to be 62.3° and 47.8°. What is the height of the balloon above the ground?

Solution 3.
Step 1. Read the problem.
Step 2. Draw a sketch.
Step 3. Identify and **label** the known and unknown quantities. Label the triangle ABC with the balloon at B. Then $A = 62.3°$, $C = 47.8°$ and $b = 500$. Since $A + B + C = 180°$, $B = 69.9°$. We want to find the height of the triangle, which we will label h.

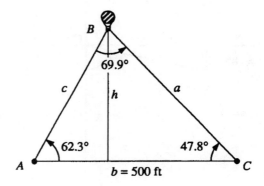

Figure 4.14

Step 4. Write equations relating the known and unknown quantities. The height h is related to side a by the equation

$$\frac{h}{a} = \sin C, \text{ or } h = a \sin 47.8°.$$

Since three angles and a side of triangle ABC are known, an equation for side a comes from the Law of Sines.

$$\frac{a}{\sin A} = \frac{b}{\sin B} \text{ or } \frac{a}{\sin 62.3°} = \frac{500}{\sin 69.9°}$$

Step 5. Solve the equations. Solve the second equation for a to find

$$a = \frac{500 \sin 62.3°}{\sin 69.9°} = 471.4.$$

Use this value for a in the first equation to find

$$h = a \sin 47.8° = 471.4 \sin 47.8° = 349.2 \text{ feet.}$$

120

Example 4. Denver is 903 miles west and 185 miles south of Chicago. What is the bearing of Denver from Chicago?

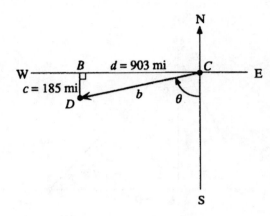

Figure 4.15

Solution 4.
Step 1. Read
Step 2. Draw
Step 3. Identify and **label** the known and unknown quantities. Chicago is at vertex C. Denver is at vertex D. Angle B is a right angle. To give the bearing we must find angle θ.

Step 4. Write equations. We know that
$$\theta = 90° - C.$$
This is a right triangle so an equation for angle C comes from trigonometric ratios.

$$\tan C = \frac{185}{903}$$

Step 5. Solve the equations.

$$\tan C = \frac{185}{903} = 0.20487$$

$$C = 11.6°$$

$$\theta = 90° - 11.6° = 78.4°$$

The bearing of Denver from Chicago is S 78.4° W.

Example 5. The lengths of two sides of a parallelogram are 5 inches and 7 inches. The length of its longer diagonal is 10 inches. What is the area of the parallelogram?

Solution 5.
Step 1. Read
Step 2. Draw
Step 3. Identify and **label**
For a parallelogram,

$$\text{area} = (\text{base}) \cdot (\text{height}).$$

The base of this parallelogram is $c = 7$. To find the area, we must find its height h.

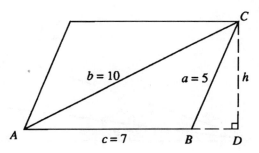

Figure 4.16

Step 4. Write equations. Since h is one leg of a right triangle ADC,

$$\sin A = \frac{h}{10}.$$

Three sides of triangle ABC are known and an equation for angle A comes from the Law of Cosines.

$$a^2 = b^2 + c^2 - 2bc \cos A$$

or

$$(5)^2 = (10)^2 + (7)^2 - 2(10)(7) \cos A$$

Step 5. Solve the equations. From the second equation

$$25 = 100 + 49 - 140 \cos A,$$

$$\cos A = \frac{124}{140} \text{ and}$$
$$A = 27.66°.$$

Use this value for A in the first equation to find
$$h = 10 \sin A = 10 \sin 27.66° = 4.64.$$

The area of the parallelogram is
$$\text{area} = (\text{base}) \cdot (\text{height}) = 7(4.64) = 32.48 \text{ square inches}$$

Example 6. A vertical tower 150 feet tall is situated on the side of a hill. At a point 650 feet down the hill the angle between the side of the hill and the line of sight to the top of the tower is 12.5°. What is the angle between the side of the hill and the horizontal?

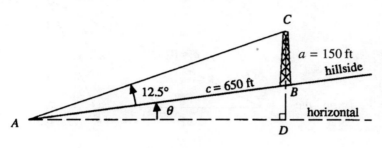

Solution 6.

Step 1. Read

Step 2. Draw

Step 3. Identify and **label**

We are to find angle θ.

Step 4. Write equations.

From the sketch,
$$\angle DAC = \theta + 12.5°.$$
Since triangle ADC is a right triangle,
$$(\theta + 12.5°) + C = 90° \quad \text{and}$$
$$\theta = 90° - 12.5 - C.$$

Figure 4.17

Two sides and an opposite angle of triangle ABC are known, so an equation for angle C comes from the Law of Sines.

$$\frac{c}{\sin C} = \frac{a}{\sin A}$$

or

$$\frac{650}{\sin C} = \frac{150}{\sin (12.5°)}$$

Step 5. Solve

$$\sin C = \frac{650 \sin (12.5°)}{150} = 0.9379.$$

$$C = 69.7° \quad \text{or} \quad 110.3°$$

It is evident from the physical configuration that C must be an acute angle, so

$$C = 69.7°.$$

The angle between the side of the hill and the horizontal is
$$\theta = 90° - 12.5° - 69.7° = 7.8°.$$

Practice Problems

1. In order to measure the height of a cloud a meteorologist beams a spotlight vertically upward. The meteorologist stands 250 meters from the spotlight. From this position she measures the angle of elevation of the spot on the cloud to be 70°. How high is the cloud?

2. A merchant ship at its dock in New York was observed to subtend an angle of 6° from the window of an office 2000 feet from the bow and 3000 feet from the stern of the ship. How long is the ship?

3. Smith and Jones, who are 4 miles apart along a straight east-west highway, see a monument some distance from the road. Smith observes the monument north and east of his location and measures the angle from the road to the monument as 44°. Jones observes the monument north and west of his location and measures the angle from the monument to the road as 37°.
(a) How far is Smith from the monument?
(b) How far is Jones from the monument?
(c) How far is the monument from the road?

4. What is the angle of elevation of the sun when a tower 45 feet high and situated on level ground casts a shadow 36 feet long?

5. The captain of a ship determines his position to be 80 miles from port and 48 miles from a lighthouse. The distance between the lighthouse and the port is 57 miles. What is the degree measure from the ship of the angle from the lighthouse to the port?

6. A vertical tower is braced by two cables which are anchored to the ground at the same point. At this point, the angle between the cables is 15°. The first cable is 150 feet long and extends to the top of the tower. The second cable extends to a point 50 feet below the top of the tower. How long is the second cable?

7. Two adjacent sides of a parallelogram are 22 inches and 28 inches in length. The angle between them is 40°. What is the area of the parallelogram?

8. An aircraft leaves an airport and flies 200 miles in direction S 40° E.
(a) How far east of the airport is the aircraft at that time?
(b) How far south of the airport is the aircraft at that time?

9. The angle of elevation from a point on level ground to the top of a tree is 38.5°. From a second point 60 feet closer to the tree the angle of elevation is 57.2°. How tall is the tree?

10. A road going straight up a hill makes a 7° angle with the horizontal. A vertical utility pole 43 feet high stands beside the road. A cable reaches from the top of the pole to a point 54 feet uphill from the pole. How long is the cable?

11. A surveyor determines the following information for a triangular shaped piece of property. One side has length 257.9 feet. A second side has length 185.1 feet. The angle opposite the 185.1 foot side measures 38.3°. How long is the third side of the property?

12. A bullet is discovered embedded in the wall of a room, with an entry point 3 meters above the floor. The path of the bullet in the wall is inclined upward at an angle of 32°. If the gun was fired from 1 meter above the floor, how far was the gun from the wall when it was fired? (Assume that the bullet traveled in a straight line.)

13. Two teams of engineering students are measuring their distance from the summit of a mountain by reflecting a laser beam off the summit. Both teams are positioned directly east of the peak at the same elevation. The first team determines that its straight-line distance from the summit is 15871 meters. The second team, positioned 2000 meters west of the first team, determines that the angle of elevation of the summit is 9.7°. What is the distance between the second team and the summit?

14. In a certain solar home the south side of the roof makes an angle of 75° with the horizontal and measures 16 feet from eaves to ridge. The north side of the roof measures 30 feet from eaves to ridge. The eaves on the two sides of the structure are at the same height. How wide is the house?

15. A ship is steaming due north at 17 miles per hour. At 12 o'clock noon the captain sights a lighthouse at bearing N 45° W from the ship. At 1 p.m. the bearing of the lighthouse from the ship is N 60° W. How far is the ship from the lighthouse at 1 p.m.?

16. From the top of a tower 30 meters above the level of a river and 15 meters from the edge of the river, the angle of depression of a point directly across the river on the opposite bank is 17°. How wide is the river?

17. An airplane race is to be flown around a triangular course. The first leg of the race is flown due east and is 152 miles long. The second and third legs lie to the north of the first leg. They are 123 miles and 105 miles long, respectively.
(a) What is the bearing of the second leg of the course?
(b) What is the bearing of the third leg of the course?

18. Each of two surveyors is located 200 feet from a marker. The angle between the marker and either surveyor, as measured by the other surveyor, is 36°. How far apart are the surveyors?

19. A certain building is 63 feet wide. The rafters on one side of the roof are 32 feet long (measured from eaves to ridge). Those on the other side are 59 feet long. The eaves on the two sides of the building are at the same height.
(a) What angle do the 32 foot rafters make with the horizontal?
(b) What angle do the 59 foot rafters make with the horizontal?

20. Two boats leave an island at the same time. The first sails 20 miles per hour with bearing N 30° W from the island. The second sails 25 miles per hour with bearing S 20° W from the island. After two hours, how far apart are the boats?

21. A guy wire 83 feet long is attached to the top of a vertical pole 57 feet high. What angle does the wire make with the pole?

22. A tower that tilts at an angle of 5° from vertical, directly away from the sun, casts a shadow 50 meters long when the angle of elevation of the sun is 58°. What is the distance from the bottom to the top of the tower? Assume the tower stands on level ground.

23. A level highway runs directly away from a cliff. From the top of the cliff the straight line distance to mile marker 123 on the highway is 0.22 miles and the distance to mile marker 124 is 1.14 miles. From the top of the cliff, what is the angle of depression of mile marker 124?

24. A ladder 24 feet long leaning against a vertical wall makes a 36° angle with the level ground.
(a) How high is the top of the ladder?
(b) How far is the foot of the ladder from the wall?

25. A large boulder and a small tree are on opposite sides of a pond. A surveyor establishes that she is at a point 92 feet from the boulder and 128 feet from the tree. From this point she measures the angle from the boulder to the tree as 37°. What is the distance between the boulder and the tree?

26. The longest diagonal of a certain parallelogram is 10 inches long. At one end the diagonal makes angles of 33° and 25° with the sides of the parallelogram. How long are the sides of the parallelogram?

27. A radar station is located 38 miles from an airport in direction N 12° E. An aircraft is approaching the airport from due east. The aircraft is 51 miles from the radar station. Approximately how far is the aircraft from the airport?

28. As an airplane ascends its path makes a 60° angle with the runway. How many feet does the plane rise while it travels 1200 feet in the air?

29. A triangular sail is made from nylon. Its three edges are 6 feet, 13.5 feet and 15.1 feet long, respectively. What is the area of the sail?

30. Alamogordo is 200 miles due south of Santa Fe. A pilot flying from Alamogordo to Santa Fe flies the first 120 miles off course with bearing N 10.3° W. How far is he from Santa Fe at that time?

UNIT 4
Sample Examination 1

1. In (possibly oblique) triangle ABC, $a = 24.5$, $b = 18.6$ and $c = 26.4$. Which one of the following might be angle B to the nearest tenth of a degree? There is at least one correct response. Choose all correct responses.

A) 37.6°   **B)** 38.3°   **C)** 39.6°   **D)** 42.7°   **E)** 45.5°

2. What is the (approximate) perimeter of (possibly oblique) triangle ABC, where $a = 132$, $b = 224$ and $C = 28°$? There is at least one correct response. Choose all correct responses.

A) 476   **B)** 480   **C)** 484   **D)** 489   **E)** 494

3. To the nearest unit, what is the length of side b of triangle ABC, where $A = 75°$, $B = 38°$ and $c = 321$? There is at least one correct response. Choose all correct responses.

A) 205   **B)** 215   **C)** 305   **D)** 410   **E)** 480

4. A crane at a loading dock is built so the base of its boom is on the loading dock. The boom is straight, 80 feet long and forms a 75° angle with the horizontal plane. A crate being lifted from the dock hangs 10 feet below the end of the boom. To the nearest foot, how far above the dock is the top of the crate?

A) 59 feet   **B)** 63 feet   **C)** 67 feet   **D)** 77 feet   **E)** 89 feet

5. At Riverside Shooting Gallery each rifle has a range of 100 feet. A mechanical target travels a straight line path across the back of the gallery. At the beginning of its path, the target is 125 feet from a rifleman and the angle between the path of the target and the path between the rifleman and the target is 40°. At what distance from the beginning of its path will the target first be within the range of the rifle?

A) 36.3 feet   **B)** 35.7 feet   **C)** 35.1 feet   **D)** 34.5 feet   **E)** 33.9 feet

UNIT 4
Sample Examination 2

1. In (possibly oblique) triangle ABC, $a = 62.5$, $b = 51.5$ and $B = 40°$. Which one of the following might be angle A to the nearest degree? There is at least one correct response. Choose all correct responses.

A) 51°   B) 72°   C) 134°   D) 129°   E) 139°

2. In (possibly oblique) triangle ABC, $a = 6.34$, $b = 7.30$ and $c = 9.98$. Which one of the following might be angle A to the nearest tenth of a degree? There is at least one correct response. Choose all correct responses.

A) 34.8°   B) 35.7°   C) 38.3°   D) 38.5°   E) 39.3°

3. Two sides of a (possibly oblique) triangle have lengths 120 and 270. The included angle measures 118°. What is the length of the third side, rounded to the nearest unit?

A) 299   B) 310   C) 325   D) 343   E) 372

4. One hole on a certain golf course is 340 yards from the tee. A golfer slices his drive (hits the ball to the right of the correct direction) at an angle of 10°. Pacing the distance to the ball, he estimates his drive was 205 yards. Assuming his estimate is correct, how far is the ball from the hole?

A) 143 yards   B) 146 yards   C) 153 yards   D) 161 yards   E) 164 yards

5. The distance along the Bower's roof from the ridge to the top of the wall of the house is 12.94 feet. The roof makes an angle of 18° with the attic floor. How high above the floor does the roof rise?

A) 3.0 feet   B) 3.2 feet   C) 3.4 feet   D) 3.7 feet   E) 4.0 feet

---

UNIT 5

GRAPHING TRIGONOMETRIC FUNCTIONS

Introduction

In Units 1 and 2 we examined the sine, cosine, tangent, cosecant, secant and cotangent as *functions*. Our principal concerns were evaluating these functions for specific angles θ and determining the angles for which the trigonometric functions have specific values.

In Units 3 and 4 we turned our attention to triangles and used the sine, cosine and tangent to describe the relationships among the sides and angles of triangles. In these Units we used the trigonometric functions as tools to study triangles and did not study the trigonometric functions themselves.

In Unit 5 we return to the study of the sine, cosine, tangent, cosecant, secant and cotangent as functions. We will re-examine the difficulties of evaluating the trigonometric functions for arbitrary angles from their definitions and use special properties of right triangles involving the angle $\frac{\pi}{4}$ $(= 45°)$, $\frac{\pi}{6}$ $(= 30°)$ and $\frac{\pi}{3}$ $(= 60°)$ to find the values of the trigonometric functions exactly for these special angles. These values are especially useful for graphing the trigonometric functions. We will study the periodic nature of these functions and use their periodicity to obtain their graphs. Finally, we will investigate how the graphs of the functions $y = A \sin B\theta$ and $y = A \cos B\theta$ are related to the graphs of $y = \sin \theta$ and $y = \cos \theta$.

UNIT 5
GRAPHING TRIGONOMETRIC FUNCTIONS

Objective 5.1

State the values of the trigonometric functions for the special angles $\dfrac{\pi}{4}$ $(= 45°)$, $\dfrac{\pi}{6}$ $(= 30°)$ and $\dfrac{\pi}{3}$ $(= 60°)$.

Objective 5.2

Sketch the graphs of the functions $y = \sin\theta$ and $y = \cos\theta$ (θ in radians), and state their amplitudes and periods.

Objective 5.3

(a) State the amplitude and the period of functions of the form $y = A\sin B\theta$ and $y = A\cos B\theta$ ($A > 0$, $B > 0$, θ in radians) and sketch their graphs.

(b) Given the graph of a function of the form $y = A\sin B\theta$ or $y = A\cos B\theta$, state the amplitude and the period of the function and determine the equation for the function.

Objective 5.4

Sketch the graphs of the tangent, cotangent, cosecant and secant functions, and state their periods.

GRAPHING TRIGONOMETRIC FUNCTIONS

Objective 5.1

State the values of the trigonometric functions for the special angles $\dfrac{\pi}{4}$ $(= 45°)$, $\dfrac{\pi}{6}$ $(= 30°)$ and $\dfrac{\pi}{3}$ $(= 60°)$.

Discussion

The sine, cosine, tangent, cosecant, secant and cotangent are called trigonometric *functions*. In mathematical language, a function consists of two parts:

- a collection **D** of mathematical objects (usually, but not always, numbers) called the domain of the function, and
- a procedure for determining exactly one number from each object in the domain.

The trigonometric functions are functions in this mathematical sense. Think first about the sine and the cosine. The domain of these functions is the set of all angles. The first step of a procedure for determining a number from an angle θ is to introduce a coordinate system so that θ is in standard position. Next, choose a point (a, b) (other than $(0, 0)$) on the terminal side of θ. The number determined from θ by the sine function is $\sin \theta = \dfrac{b}{\sqrt{a^2 + b^2}}$ and the number determined from θ by the cosine function is $\cos \theta = \dfrac{a}{\sqrt{a^2 + b^2}}$.

The domain of the tangent and the secant is the set of all angles that do not have terminal side along the y-axis. The number determined from θ by the tangent function is $\tan \theta = \dfrac{b}{a}$ and the number determined by the secant function is $\sec \theta = \dfrac{\sqrt{a^2 + b^2}}{a}$, where (a, b) is a point on the terminal side of θ. When the terminal side of θ is along the y-axis, $a = 0$, so these quotients are not defined. For this reason, these angles are excluded from the domain of the tangent and secant.

The domain of the cotangent and the cosecant is the set of all angles that do not have terminal side along the x-axis. The number determined from θ by the cotangent function is $\cot \theta = \dfrac{a}{b}$ and the number determined by the cosecant function is $\csc \theta = \dfrac{\sqrt{a^2 + b^2}}{b}$, where (a, b) is a point on the terminal side of θ. When the terminal side of θ is along the x-axis, $b = 0$, so these quotients are not defined. For this reason, these angles are excluded from the domain of the cotangent and cosecant.

Angles can be specified in degrees, radians or revolutions and fractions thereof. When trigonometric functions are being thought of as *functions* (rather than as tools for solving triangles) radians are the best mode for expressing angles. We will use radians exclusively in studying graphs of trigonometric functions in this Unit.

It is usually very difficult to obtain exact values of the trigonometric functions for a specific angle θ from the definitions because we cannot find a specific point on the terminal side of the angle. We saw in Objective 2.1 that the quadrantal angles are an exception because it is easy to identify points on terminal sides that lie along coordinate axes. The special angles $\dfrac{\pi}{4}$ $(= 45°)$, $\dfrac{\pi}{6}$ $(= 30°)$ and $\dfrac{\pi}{3}$ $(= 60°)$ are also exceptions. By using trigonometric ratios to describe the special relation-ships among sides and angles of right triangles that include one of these acute angles, we can obtain exact values of the trigonometric functions for these special angles.

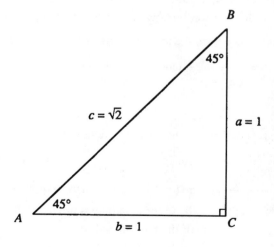

For the 45° angle consider the 45° − 45° right triangle *ACB* shown in Figure 5.1. Since this is an isosceles triangle, the two sides opposite the 45° angles are of equal length. Assume these sides have length $a = b = 1$. By the Pythagorean Theorem the hypot-enuse has length

$$c = \sqrt{a^2 + b^2} = \sqrt{1^2 + 1^2} = \sqrt{2}.$$

Figure 5.1

As in Objective 3.1, we can use trigonometric ratios to relate the sides and angles of this right triangle. We obtain

$$\sin\frac{\pi}{4} = \sin 45° = \frac{\text{side opposite}}{\text{hypotenuse}} = \frac{1}{\sqrt{2}} = \frac{\sqrt{2}}{2},$$

$$\cos\frac{\pi}{4} = \cos 45° = \frac{\text{side adjacent}}{\text{hypotenuse}} = \frac{1}{\sqrt{2}} = \frac{\sqrt{2}}{2} \quad\text{and}$$

$$\tan\frac{\pi}{4} = \tan 45° = \frac{\text{side opposite}}{\text{side adjacent}} = \frac{1}{1} = 1.$$

To find the values of the trigonometric function for 30° and 60° angles, construct an equilateral triangle ABC with sides of length 2 as in Figure 5.2.

In this triangle $a = b = c = 2$ and $A = B = C = 60°$. From vertex B construct a perpendicular BD to the opposite side. This line divides angle B into two 30° angles. It divides the base of the triangle into two equal parts, each of length 1. It divides the equilateral triangle ABC into two 30° − 60° right triangles ADB and CDB.

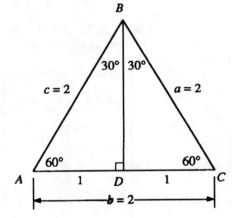

Figure 5.2

Now, concentrate on the right triangle ADB which is the left half of the original triangle in Figure 5.2. Right triangle ADB is shown in Figure 5.3. The height h of right triangle ADB is found by the Pythagorean Theorem:

$$1^2 + h^2 = 2^2$$

so

$$h = \sqrt{3}.$$

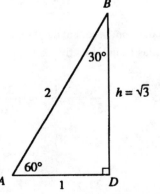

Figure 5.3

From the trigonometric ratios relating the sides and angles of right triangle ADB (recall Objective 3.1) we have

$$\sin \frac{\pi}{3} = \sin 60° = \frac{\sqrt{3}}{2}, \qquad \sin \frac{\pi}{6} = \sin 30° = \frac{1}{2},$$

$$\cos \frac{\pi}{3} = \cos 60° = \frac{1}{2}, \qquad \cos \frac{\pi}{6} = \cos 30° = \frac{\sqrt{3}}{2},$$

$$\tan \frac{\pi}{3} = \tan 60° = \frac{\sqrt{3}}{1}, \qquad \tan \frac{\pi}{6} = \tan 30° = \frac{1}{\sqrt{3}} = \frac{\sqrt{3}}{3}.$$

The values of sine, cosine and tangent for the special angles

$$\frac{\pi}{4} \; (= 45°), \; \frac{\pi}{6} \; (= 30°) \; \text{ and } \; \frac{\pi}{3} \; (= 60°)$$

are summarized in Table 5.1. The reader should memorize these function values or learn to reconstruct them quickly and easily by the procedures just discussed.

angle θ		$\sin\theta$	$\cos\theta$	$\tan\theta$
degrees	radians			
45°	$\frac{\pi}{4}$	$\frac{\sqrt{2}}{2} \; \frac{\sqrt{2}}{2}$	1	
30°	$\frac{\pi}{6}$	$\frac{1}{2}$	$\frac{\sqrt{3}}{2}$	$\frac{\sqrt{3}}{3}$
60°	$\frac{\pi}{3}$	$\frac{\sqrt{3}}{2}$	$\frac{1}{2}$	$\sqrt{3}$

Table 5.1 Values of Trigonometric Functions for Special Angles

The cosecant, secant and cotangent can be evaluated for these special angles by using the reciprocal formulas as in Objective 2.2. For example,

$$\csc 45° = \frac{1}{\sin 45°} = \frac{1}{\sqrt{2}/2} = \frac{2}{\sqrt{2}} = \sqrt{2}, \qquad \sec 30° = \frac{1}{\cos 30°} = \frac{1}{\sqrt{3}/2} = \frac{2}{\sqrt{3}} = \frac{2\sqrt{3}}{3},$$

$$\cot 60° = \frac{1}{\tan 60°} = \frac{1}{\sqrt{3}/1} = \frac{1}{\sqrt{3}} = \frac{\sqrt{3}}{3}.$$

Summary

1. In mathematical language, a function consists of two parts:
   - a collection **D** of mathematical objects (usually, but not always, numbers) called the domain of the function, and
   - a procedure for determining exactly one number from each object in the domain.

   The trigonometric functions are functions in this mathematical sense.

2. The domain of the sine and cosine functions is the set of all angles. The number determined from θ by the sine function is $\sin \theta = \dfrac{b}{\sqrt{a^2 + b^2}}$ and the number determined from θ by the cosine function is $\cos \theta = \dfrac{a}{\sqrt{a^2 + b^2}}$, where (a, b) is a point on the terminal side of θ.

   The domain of the tangent and the secant is the set of all angles that do not have terminal side along the y-axis. The number determined from θ by the tangent function is $\tan \theta = \dfrac{b}{a}$ and the number determined by the secant function is $\sec \theta = \dfrac{\sqrt{a^2 + b^2}}{a}$, where (a, b) is a point on the terminal side of θ.

   The domain of the cotangent and the cosecant is the set of all angles that do not have terminal side along the x-axis. The number determined from θ by the cotangent function is $\cot \theta = \dfrac{a}{b}$ and the number determined by the cosecant function is $\csc \theta = \dfrac{\sqrt{a^2 + b^2}}{b}$, where (a, b) is a point on the terminal side of θ.

3. By using trigonometric ratios to describe the special relationships among sides and angles of right triangles that include one of these acute angles, we obtain the following exact values of the sine, cosine and tangent functions for the special angles $\dfrac{\pi}{4}$ ($= 45°$), $\dfrac{\pi}{6}$ ($= 30°$) and $\dfrac{\pi}{3}$ ($= 60°$).

angle θ		$\sin \theta$	$\cos \theta$	$\tan \theta$
degrees	radians			
45°	$\dfrac{\pi}{4}$	$\dfrac{\sqrt{2}}{2}$	$\dfrac{\sqrt{2}}{2}$	1
30°	$\dfrac{\pi}{6}$	$\dfrac{1}{2}$	$\dfrac{\sqrt{3}}{2}$	$\dfrac{\sqrt{3}}{3}$
60°	$\dfrac{\pi}{3}$	$\dfrac{\sqrt{3}}{2}$	$\dfrac{1}{2}$	$\sqrt{3}$

The values of the cosecant, secant and cotangent for these special angles can be found by using the reciprocal formulas.

Examples

Example 1. Find $2 \sin 45° \sec 45°$.

> **Solution 1.** From Table N5.1 we have $\sin 45° = \dfrac{\sqrt{2}}{2}$ and $\sec 45° = \dfrac{1}{\cos 45°} = \dfrac{2}{\sqrt{2}}$.
> Thus,
>
> $$2 \sin 45° \sec 45° = 2 \cdot \frac{\sqrt{2}}{2} \cdot \frac{2}{\sqrt{2}} = 2.$$

Example 2. Find $\cos \dfrac{\pi}{6} - \tan \dfrac{\pi}{3} \csc \dfrac{\pi}{3}$.

> **Solution 2.** We have $\cos \dfrac{\pi}{6} = \dfrac{\sqrt{3}}{2}$, $\tan \dfrac{\pi}{3} = \sqrt{3}$ and $\csc \dfrac{\pi}{3} = \dfrac{1}{\sin \pi/3} = \dfrac{2}{\sqrt{3}}$.
>
> Thus,
> $$\cos \frac{\pi}{6} - \tan \frac{\pi}{3} \csc \frac{\pi}{3} = \frac{\sqrt{3}}{2} - \sqrt{3} \cdot \frac{2}{\sqrt{3}}$$
>
> $$= \frac{\sqrt{3}}{2} - 2 = \frac{\sqrt{3} - 4}{2}.$$

Practice Problems

Evaluate the following expressions without using a calculator.

1. $\dfrac{1 - \cos 60°}{\sin 60°}$

2. $\sec \dfrac{\pi}{4} \tan \dfrac{\pi}{4}$

3. $\dfrac{2 \tan 45°}{2 - \sec^2 45°}$

4. $\csc \dfrac{\pi}{6} + \cot \dfrac{\pi}{6}$

5. $2 \cos \dfrac{\pi}{3} \sin \dfrac{\pi}{4}$

6. $\cos^2 45° - \sin^2 30°$

7. $\dfrac{\csc^2 60° - 2}{2 \cot 60°}$

8. $\dfrac{\tan \pi/6 + \tan \pi/3}{1 - \tan \pi/6 \ \tan \pi/3}$

9. $\cos \dfrac{\pi}{4} \cos \dfrac{\pi}{6} - \sin \dfrac{\pi}{4} \sin \dfrac{\pi}{6}$

10. $1 - 2 \sin^2 45°$

11. $\csc^2 30° - \cot^2 30°$

12. $\sin \dfrac{\pi}{3} \cos \dfrac{\pi}{6} - \cos \dfrac{\pi}{3} \sin \dfrac{\pi}{6}$

13. $\sec^2 \dfrac{\pi}{3} + \tan^2 \dfrac{\pi}{3}$

14. $\dfrac{\sec^2 30°}{1 + \tan 30°}$

15. $\dfrac{\csc^2 \pi/4}{\sqrt{1 + \cot^2 \pi/4}}$

16. $\sec^2 \dfrac{\pi}{6} + \csc^2 \dfrac{\pi}{3}$

17. $2 \csc 45° \cot 45°$

18. $\dfrac{1 - \sec 60°}{1 + \cos 60°}$

19. What is the exact numerical value of $\dfrac{\cot^2 \pi/3 - 1}{2 \cot \pi/3}$?

(A) $\dfrac{\sqrt{3}}{2}$ (B) $\dfrac{\sqrt{3} - 3}{6}$ (C) $\dfrac{\sqrt{3}}{3}$ (D) $\dfrac{\sqrt{3} - 1}{2}$ (E) $\dfrac{-\sqrt{3}}{3}$

20. What is the exact numerical value of $(1 + \cos 30°) \tan 60°$?

(A) $\dfrac{2\sqrt{3} + \sqrt{6}}{2}$ (B) $\dfrac{2\sqrt{3} + 3}{2}$ (C) $\dfrac{3\sqrt{3}}{2}$ (D) $\dfrac{2\sqrt{3} + \sqrt{6}}{4}$ (E) $\dfrac{2\sqrt{3} + 3}{6}$

Objective 5.2

Sketch the graphs of the functions $y = \sin \theta$ and $y = \cos \theta$ (θ in radians), and state their amplitudes and periods.

Discussion

The graph of a function $y = f(x)$ is the collection of points in a coordinate plane whose coordinates (a, b) satisfy $b = f(a)$. The graph, which is often a curve in the plane, gives us a way to visualize a function in its entirety. Our goal in this Discussion is to acquire a global view of the sine and cosine functions by generating their graphs.

We immediately encounter a minor difficulty graphing any trigonometric function. The coordinates of points on a graph are pairs of numbers (a, b), where the first *number* comes from the *domain* of the function. But the domains of the trigonometric functions are *angles*, not numbers. How could a point (θ, b) whose first coordinate θ is a geometric object be plotted? To avoid this difficulty, we identify angles with their measures and plot the measure of the angle (a number) instead of wondering how to plot an angle. Having made this identification, we must decide which mode of angular measure to use. Calculus is a tool for studying functions. Radians are more compatible with the processes of calculus than other modes of angular measure. As a result, radians are used almost exclusively in the study of the trigonometric functions.

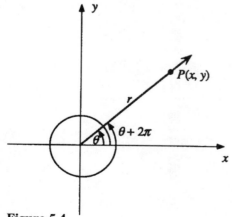

We say a function is **periodic** to mean that it repeats its values at regular intervals. The trigonometric functions are periodic because they repeat their values every 2π radians ($360°$). To see that the trigonometric functions are periodic, consider the angle θ and the angle $\theta + 2\pi$ as shown in Figure 5.4. These two angles share the same terminal side. To evaluate the trigonometric functions at θ and and at $\theta + 2\pi$, choose a point $P(x, y)$ on the common terminal side of these angles. By expressing the values of the trigonometric functions in terms of this point, we see that these functions have the same values at θ and $\theta + 2\pi$. That is,

Figure 5.4

(1)

$$\sin \theta = \sin (\theta + 2\pi) = \frac{y}{r}, \qquad \csc \theta = \csc (\theta + 2\pi) = \frac{r}{y},$$

$$\cos \theta = \cos (\theta + 2\pi) = \frac{x}{r}, \qquad \sec \theta = \sec (\theta + 2\pi) = \frac{r}{x},$$

$$\tan \theta = \tan (\theta + 2\pi) = \frac{y}{x}, \qquad \cot \theta = \cot (\theta + 2\pi) = \frac{x}{y}.$$

By the same argument, trigonometric functions have the same values for the angles

$$\theta, \quad \theta + 2\pi, \quad \theta + 4\pi, \quad \theta + 6\pi, \quad \dots$$

and

$$\theta - 2\pi, \quad \theta - 4\pi, \quad \theta - 6\pi, \quad \dots \ .$$

We conclude the following:

The trigonometric functions repeat their values every 2π units.

For example, as you can verify with your calculator,

$$\sin \frac{\pi}{6} = \sin \left(\frac{\pi}{6} + 2\pi\right) = \sin \left(\frac{\pi}{6} + 4\pi\right) = \dots$$

$$= \sin \left(\frac{\pi}{6} - 2\pi\right) = \sin \left(\frac{\pi}{6} - 4\pi\right) = \dots$$

$$= \frac{1}{2}.$$

Because the trigonometric functions are periodic and repeat their values every 2π radians, sections of their graphs over successive intervals of length 2π are identical. For example, the graph of $y = \sin \theta$ over the interval $0 \le \theta \le 2\pi$ looks exactly like the graph over the interval $2\pi \le \theta \le 4\pi$.

The direct approach to constructing the graph of a function is to first evaluate the function at a sample of values from across its domain, then to plot the corresponding points in the coordinate plane and, finally, to draw a smooth curve through the points plotted. Since the trigonometric functions repeat their values over intervals of length 2π, we can obtain the entire graph of a trigonometric function by constructing the graph carefully on the interval from 0 to 2π and then duplicating this portion of the graph over successive intervals of length 2π. The first step in graphing the sine and cosine functions by this direct approach is to make a table of exact values of the sine and cosine at the quadrantal angles and the angles between 0 and 2π which have the special angles studied in Objective 5.1 as their reference angles. Table 5.2 lists values for angles which are multiples of $\frac{\pi}{6}$.

angle θ		$\sin\theta$	$\cos\theta$
radians	degrees		
0	$0°$	0	1
$\dfrac{\pi}{6}$	$30°$	$\dfrac{1}{2}=.5$	$\dfrac{\sqrt{3}}{2}=.87$
$\dfrac{\pi}{3}$	$60°$	$\dfrac{\sqrt{3}}{2}=.87$	$\dfrac{1}{2}=.5$
$\dfrac{\pi}{2}$	$90°$	1	0
$\dfrac{2\pi}{3}$	$120°$	$\dfrac{\sqrt{3}}{2}=.87$	$-\dfrac{1}{2}=-.5$
$\dfrac{5\pi}{6}$	$150°$	$\dfrac{1}{2}=.5$	$-\dfrac{\sqrt{3}}{2}=-.87$
π	$180°$	0	-1
$\dfrac{7\pi}{6}$	$210°$	$-\dfrac{1}{2}=-.5$	$-\dfrac{\sqrt{3}}{2}=-.87$
$\dfrac{4\pi}{3}$	$240°$	$-\dfrac{\sqrt{3}}{2}=-.87$	$-\dfrac{1}{2}=-.5$
$\dfrac{3\pi}{2}$	$270°$	-1	0
$\dfrac{5\pi}{3}$	$300°$	$-\dfrac{\sqrt{3}}{2}=-.87$	$\dfrac{1}{2}=.5$
$\dfrac{11\pi}{6}$	$330°$	$-\dfrac{1}{2}=-.5$	$\dfrac{\sqrt{3}}{2}=.87$
2π	$360°$	0	1

Table 5.2

The values for $\theta = \dfrac{\pi}{6}$ and $\theta = \dfrac{\pi}{3}$ are used to compute entries for angles having these reference angles. For example, the angle $\theta = \dfrac{7\pi}{6}$ has $\theta' = \dfrac{7\pi}{6} - \pi = \dfrac{\pi}{6}$ as reference angle. Since the sine is negative for quadrant III angles,

$$\sin\frac{7\pi}{6} = -\sin\frac{\pi}{6} = -\frac{1}{2}.$$

The second step is to plot the points determined by Table 5.2 and draw a smooth curve through them. The resulting graphs are shown in Figures 5.5 and 5.6.

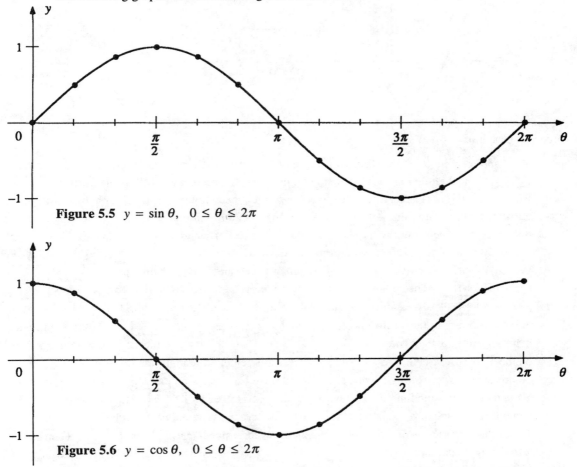

Figure 5.5 $y = \sin \theta, \quad 0 \le \theta \le 2\pi$

Figure 5.6 $y = \cos \theta, \quad 0 \le \theta \le 2\pi$

The third step is to extend to graphs all values of θ. Since the sine and cosine functions repeat their values every 2π units, the curves for $0 \le \theta \le 2\pi$ are duplicated on the interval $2\pi \le \theta \le 4\pi$; they are again duplicated on the interval $4\pi \le \theta \le 6\pi$; and so on. The extended graphs are shown in Figures 5.7 and 5.8.

The graphs of $y = \sin \theta$ and $y = \cos \theta$ can be generated easily on an advanced scientific or graphics calculator. The procedures for graphing will be different for various makes and models of graphics calculators. A brief description of the procedure for graphing the sine and cosine functions on the Texas Instruments TI-82® follows. Consult the operator's manual to learn the procedure for the particular machine you are using or for a more detailed discussion of graphing with the Texas InstrumentsTI-82®.

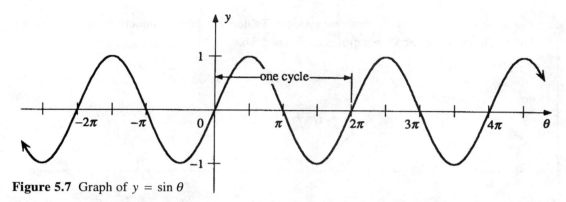

Figure 5.7 Graph of $y = \sin \theta$

To graph $y = \sin \theta$ on the Texas Instruments TI-82® , first access the mode menu by pressing
[MODE] . Select radians as the unit of angular measure by scrolling to **Radian** and pressing [ENTER] .
Before leaving this screen, set the machine to connect points on the graph by scrolling to
Connected and pressing [ENTER] . Next, define the coordinate grid on which the graph will be
displayed by specifying the largest and smallest values of x and y to be displayed. To establish
$-\pi$ to π as the interval over which the function will be graphed, press [WINDOW] to display the
WINDOW variables edit screen. Move the cursor to the line to define **Xmin**, press [(−)] , [2nd] ,
[π] and [ENTER] . The cursor will automatically move to the line to define **Xmax**. Press [2nd] ,
[π] and [ENTER] . To establish the interval -2 to 2 as the range of y values that will be displayed,
move the cursor to the **Ymin** line, then press [(−)] , [2] and [ENTER] . The cursor will automatically move to the **Ymax** line. Press [2] and [ENTER] . Next, define the function to be graphed
on the Y = edit screen. Press [Y=] to access this screen. Clear all functions previously entered
on this screen by scrolling to each function displayed and pressing [CLEAR] . Position the cursor at
Y1 and enter the sine function by pressing [SIN] followed by [X,T,Θ] . Finally, press [GRAPH] to
generate the graph of $y = \sin \theta$ on the screen. The screen shows only the graph between $-\pi$ and π .
To see that this section of the graph is repeated on successive intervals, press [TRACE] . Then use the
left and right cursor move keys [◄] and [►] to pan across the graph. To generate the graph
of $y = \cos \theta$, follow the same procedure, but enter the cosine function on the Y = edit screen.

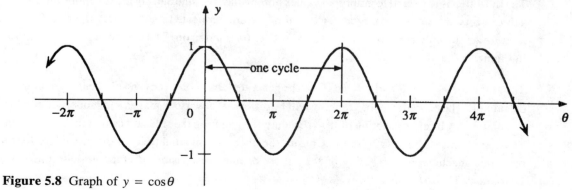

Figure 5.8 Graph of $y = \cos \theta$

As we have seen, the graphs of the sine and cosine repeat their values over intervals of length 2π. Their entire graphs consist of duplicates of the part of the graph from 0 to 2π on successive intervals of length 2π. The portion of the graph for $0 \leq \theta \leq 2\pi$ is called a **cycle** of the sine or cosine. An interval of length less than 2π does not give a full cycle of the sine or cosine. Because the sine and cosine repeat, we say they are **periodic functions.** Because a cycle occurs over an interval of length 2π, and no shorter interval, the number 2π is called their **period**.

The graphs of the sine and cosine oscillate above and below the horizontal axis, ranging from one unit above it to one unit below it. The maximum distance between the graph of the sine or cosine and the horizontal axis is called the **amplitude** of the functions. Thus, the functions $y = \sin \theta$ and $y = \cos \theta$ have amplitude 1. In general, the amplitude of a periodic function is half of the difference between its maximum and minimum values.

In this Discussion the sine and cosine were graphed as functions of the angle θ in radians. To obtain their graphs as functions of θ in degrees, simply relabel the units on the horizontal axis accordingly.

Summary

1. The values of the trigonometric functions for any angle θ in standard position are determined by a point on the terminal side of θ. Since θ (radians) and $\theta + 2\pi$ have the same terminal side, the trigonometric functions have the same values for $\theta + 2\pi$ as they have for θ. In other words, the trigonometric functions repeat their values every 2π units.

2. To construct the graph of $y = \sin \theta$ or $y = \cos \theta$ manually, sketch the graph of the function on the interval from 0 to 2π by plotting points corresponding to quadrantal angles and angles which have reference angle $\dfrac{\pi}{6}$, $\dfrac{\pi}{4}$ or $\dfrac{\pi}{3}$ (no calculator needed). Extend the graph to all angles by duplicating this curve over successive intervals of length 2π.

3. To generate the graph of $y = \sin \theta$ or $y = \cos \theta$ on an advanced scientific or graphics calculator, use the graphing procedures described in the operator's manual for the machine you are using.

4. Because 2π is the smallest number such that for all angles θ, $\sin(\theta + 2\pi) = \sin \theta$ and $\cos(\theta + 2\pi) = \cos \theta$, we say that the sine and cosine are **periodic** with **period** 2π.

5. The amplitude of a periodic function is one-half of the difference between its largest and smallest values. The sine and cosine have largest value 1 and smallest value -1, so the amplitude of the sine and cosine is 1. Amplitude is undefined for periodic functions that do not have both a largest and a smallest value.

Practice Problems

1. What is the amplitude of $y = \cos\theta$? **3.** What is the period of $y = \cos\theta$?

2. What is the period of $y = \sin\theta$? **4.** What is the amplitude of $y = \sin\theta$?

5. Fill in the four tables of values for the sine and cosine functions. Express values to the nearest hundredth (only Table 5.3 requires a scientific calculator).

θ	$\sin\theta$	$\cos\theta$
0		
$\dfrac{\pi}{12}$		
$\dfrac{\pi}{6}$		
$\dfrac{\pi}{4}$		
$\dfrac{\pi}{3}$		
$\dfrac{5\pi}{12}$		

Table 5.3

θ	$\sin\theta$	$\cos\theta$
$\dfrac{\pi}{2}$		
$\dfrac{7\pi}{12}$		
$\dfrac{2\pi}{3}$		
$\dfrac{3\pi}{4}$		
$\dfrac{5\pi}{6}$		
$\dfrac{11\pi}{12}$		

Table 5.4

θ	$\sin\theta$	$\cos\theta$
π		
$\dfrac{13\pi}{12}$		
$\dfrac{7\pi}{6}$		
$\dfrac{5\pi}{4}$		
$\dfrac{4\pi}{3}$		
$\dfrac{17\pi}{12}$		

Table 5.5

θ	$\sin\theta$	$\cos\theta$
$\dfrac{3\pi}{2}$		
$\dfrac{19\pi}{12}$		
$\dfrac{5\pi}{3}$		
$\dfrac{7\pi}{4}$		
$\dfrac{11\pi}{6}$		
$\dfrac{23\pi}{12}$		
2π		

Table 5.6

6. Use the tables in Problem 5 to graph $y = \sin \theta$ *for* $0 \le \theta \le 2\pi$ in the coordinate system given in Figure 5.9.

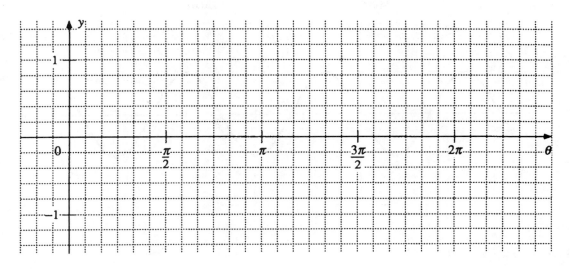

Figure 5.9

7. Use the tables in Problem 5 to graph $y = \cos \theta$ for $0 \le \theta \le 2\pi$ in the coordinate system given in Figure 5.10.

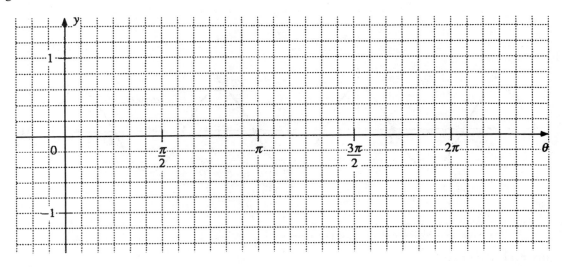

Figure 5.10

8. What are the coordinates of the points *A, B, C, D, E, F,* and *G* on the graph of $y = \sin \theta$ shown in Figure 5.11?

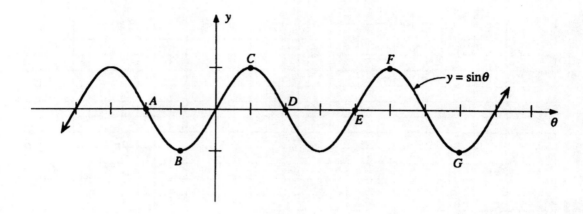

Figure 5.11 $y = \sin \theta$

9. What are the coordinates of the points *A, B, C, D, E, F,* and *G* on the graph of $y = \cos \theta$ shown in Figure 5.12?

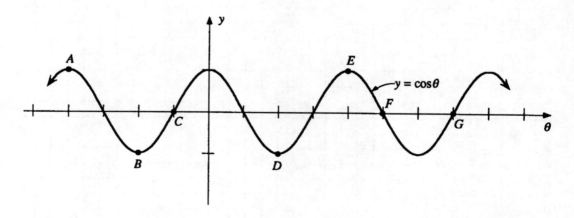

Figure 5.12 $y = \cos \theta$

10. What are the coordinates of the point P on the graph of $y = \sin \theta$ shown in Figure 5.13?

(A) $(3, -1)$ (B) $(3\pi, -2)$ (C) $\left(\dfrac{3\pi}{2}, -1\right)$ (D) $\left(\dfrac{3\pi}{4}, -2\right)$ (E) $(2\pi, -1)$

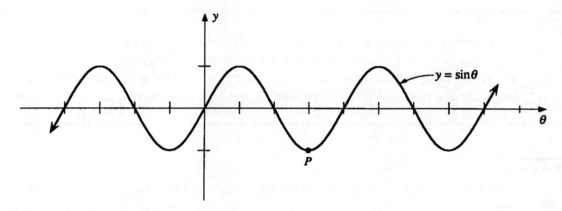

Figure 5.13 $y = \sin \theta$

11. What is the first coordinate of the point Q on the graph of $y = \cos \theta$ shown in Figure 5.14?

(A) 3   (B) 3π (C) $\dfrac{3\pi}{4}$ (D) 2π (E) $\dfrac{3\pi}{2}$

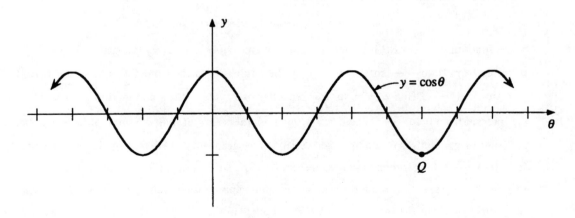

Figure 5.14 $y = \cos \theta$

Objective 5.3

(a) **State the amplitude and state the period of functions of the form** $y = A \sin B\theta$ **and** $y = A \cos B\theta$ ($A > 0$, $B > 0$, θ **in radians) and sketch their graphs.**

(b) **Given the graph of a function of the form** $y = A \sin B\theta$ **or** $y = A \cos B\theta$**, state the amplitude and state the period of the function and determine the equation for the function.**

Discussion

The functions $y = A \sin B\theta$ and $y = A \cos B\theta$, where A and B represent positive constants, are generalizations of the basic sine and cosine functions. The graphs of these more general functions are distortions of the graphs of the basic sine and cosine. The constant A has the effect of compressing or expanding the basic sine or cosine graph in the vertical direction. The constant B has a similar effect in the horizontal direction. Thus, the graphs of these more general functions resemble the graphs of the basic sine and cosine but have different periods and amplitudes.

To see how these distortions come about, consider the function $y = \sin B\theta$. The constant B affects the period of the function. For any angle θ,

$$\sin B\left(\theta + \frac{2\pi}{B}\right) = \sin\left(B\theta + 2\pi\right) = \sin B\theta$$

so the function $y = \sin B\theta$ is periodic and repeats its values over every interval of length $\frac{2\pi}{B}$. As θ varies from 0 to $\frac{2\pi}{B}$ (so $0 \leq \theta \leq \frac{2\pi}{B}$), the angle $B\theta$ varies from 0 to 2π, so the period of $y = \sin B\theta$ is $\frac{2\pi}{B}$. Hence, $y = \sin B\theta$ generates one cycle of the sine function as θ varies over the interval from 0 to $\frac{2\pi}{B}$. When $B > 1$ this interval is shorter than the interval from 0 to 2π, so the graph of $y = \sin B\theta$ is a compressed version of the graph of $y = \sin\theta$. When $B < 1$ the interval from 0 to $\frac{2\pi}{B}$ is longer than the interval from 0 to 2π, so the graph of $y = \sin B\theta$ is an expanded version of the graph of $y = \sin\theta$. Similar reasoning applies to the function $y = \sin B\theta$. Thus the functions $y = A \sin B\theta$ and $y = A \cos B\theta$ are both periodic with period $\frac{2\pi}{B}$. Since A is a constant, the functions $y = A \sin B\theta$ and $y = \cos B\theta$ are also periodic with period $\frac{2\pi}{B}$.

The role of the constant A in the functions $y = A \sin B\theta$ and $y = A \cos B\theta$ is easier to understand. It simply multiplies the values of the sine or cosine by a factor of A. For example, if $A = 2$, then the values of $\sin B\theta$ or $\cos B\theta$ are doubled, and if $A = \frac{1}{2}$, the values of $\sin B\theta$ or $\cos B\theta$ are halved. Since the sine and cosine functions oscillate between -1 and 1 and have amplitude 1, the functions $y = A \sin B\theta$ and $y = A \cos B\theta$ oscillate between $-A$ and A and have amplitude A.

The graph of $y = A \sin B\theta$ and $y = A \cos B\theta$ can be obtained by distorting the graph of $y = \sin \theta$ or $y = \cos \theta$. To obtain the graph of $y = A \sin B\theta$, begin with the cycle of the basic sine curve on the interval from . To obtain the graph of $y = A \cos B\theta$, begin with the cycle of the basic cosine curve on the interval from 0 to 2π. First, compress or expand the basic curve horizontally so this one cycle fills the interval from 0 to $\frac{2\pi}{B}$ (instead of 0 to 2π). Then, compress or expand the graph vertically so it oscillates between $-A$ and A (instead of between -1 and 1). Finally, extend the graph to all angles θ by duplicating this curve over successive intervals of length $\frac{2\pi}{B}$. The graphs in Figures 5.15 and 5.16 illustrate the combined effects of these distortions.

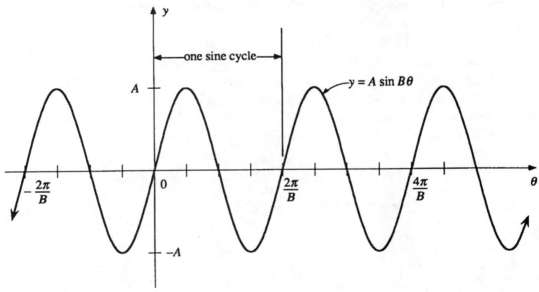

Figure 5.15 Graph of $y = A \sin B\theta$

By graphing both of the functions $y = \sin \theta$ and $y = A \sin B\theta$ (or $y = \cos \theta$ and $y = A \cos B\theta$) at the same time on an advanced scientific or graphics calculator, one can see more clearly how the graphs of the more general functions $y = A \sin B\theta$ and $y = A \cos B\theta$ are vertical and horizontal distortions of the graphs of $y = \sin \theta$ and $y = \cos \theta$.

The procedures for graphing will be different for various makes and models of graphics calculators. A brief description of the procedure for graphing functions on the Texas Instruments TI-82®

follows. Consult the operator's manual to learn the procedure for graphing functions on the particular machine you are using or for a more detailed discussion of graphing with the Texas Insturments TI-82®.

To graph $y = \sin\theta$ and $y = 2\sin 3\theta$ on the Texas Instruments TI-82®, first access the mode menu by pressing $\boxed{\text{MODE}}$. Select radians as the unit of angular measure by scrolling to **Radian** and pressing $\boxed{\text{ENTER}}$. Set the machine to connect points on the graphs by scrolling to **Connected** and pressing $\boxed{\text{ENTER}}$. Next, define the coordinate grid on which the graph will be displayed by specifying the largest and smallest values of x and y to be displayed. To establish $-\pi$ to π as the interval over which the functions will be graphed press $\boxed{\text{WINDOW}}$ to display the WINDOW variables edit screen. Move the cursor to the line to define **Xmin**, press $\boxed{\text{(-)}}$, $\boxed{\text{2nd}}$, [π] and $\boxed{\text{ENTER}}$. The cursor will automatically move to the line to define **Xmax**. Press $\boxed{\text{2nd}}$,[π] and $\boxed{\text{ENTER}}$. To establish the interval -2 to 2 as the range of y values that will be displayed, move the cursor to the **Ymin** line, press $\boxed{\text{(-)}}$, $\boxed{2}$ and $\boxed{\text{ENTER}}$. The cursor will automatically move to the **Ymax** line. Press $\boxed{2}$ and $\boxed{\text{ENTER}}$. Next, define the functions to be graphed on the Y = edit screen. Press $\boxed{\text{Y=}}$ to access this screen. Clear all functions previously entered on this screen by scrolling to each function displayed and pressing $\boxed{\text{CLEAR}}$. Position the cursor at **Y1** and enter the sine function by pressing $\boxed{\text{SIN}}$ followed by $\boxed{\text{X,T,}\Theta}$. Position the cursor at **Y2** and enter the second function by a similar sequence of key strokes. Finally, press $\boxed{\text{GRAPH}}$ to generate first the graph of $y = \sin\theta$ and then the graph of $y = 2\sin 3\theta$ on the screen. Investigate the effects of A and B by entering other functions of the form $y = A\sin B\theta$ as **Y2** and comparing the graphs. It may be necessary to adjust the size of the coordinate system on the WINDOW variables edit screen in order to see the full effect of A and B. Investigate $y = \cos\theta$ and $y = A\cos B\theta$ in the same way.

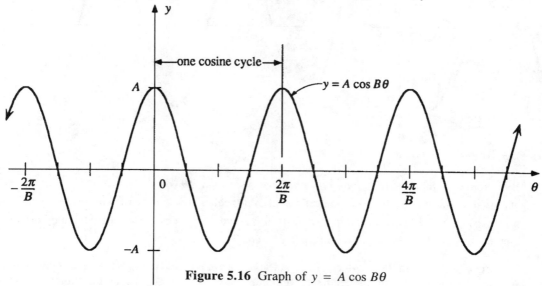

Figure 5.16 Graph of $y = A\cos B\theta$

Summary

1. The functions $y = A \sin B\theta$ and $y = A \cos B\theta$ are periodic with period $\dfrac{2\pi}{B}$ and amplitude A.

2. The amplitude of the functions $y = A \sin B\theta$ and $y = A \cos B\theta$ is the maximum distance the graph of the function oscillates above or below the horizontal axis.

3. The period of functions $y = A \sin B\theta$ and $y = A \cos B\theta$ is the length of the longest interval over which the graph of the function does not begin to repeat itself (*i.e.*, an interval which contains exactly one cycle of the function).

4. To sketch the graph of $y = A \sin B\theta$ (or $y = A \cos B\theta$) begin with the cycle of the basic sine curve (or cosine curve) on the interval $[0, 2\pi]$. First, compress or expand the basic curve horizontally so this one cycle fills the interval from 0 to $\dfrac{2\pi}{B}$. Then, compress or expand the graph vertically so it oscillates between $-A$ and A. Finally, extend the graph by duplicating this curve over successive intervals of length $\dfrac{2\pi}{B}$.

5. By graphing both of the functions $y = \sin \theta$ and $y = A \sin B\theta$ (or $y = \cos \theta$ and $y = A \cos B\theta$) at the same time on an advanced scientific or graphics calculator, one can see more clearly how the graphs of the more general functions $y = A \sin B\theta$ and $y = A \cos B\theta$ are vertical and horizontal distortions of the graphs of $y = \sin \theta$ and $y = \cos \theta$. To generate the graphs of these functions on an ASGC, use the graphing procedures described in the operator's manual for the machine you are using.

Authors' Note

As one graphs a function, it quickly becomes clear that the graph is actually a *picture* of the function. Just as a schematic drawing helps us understand how a device works, a graph helps us visualize how a function works. In the same way that drawings, paintings and photographs of an object make us aware of qualities and attributes that might otherwise have gone unnoticed, the graph of a function helps us recognize patterns and trends in the behavior of a function over its domain. The graphs of the trigonometric functions, for example, reveal a surprising beauty and symmetry in these functions.

Carl B. Boyer, in *A History of Mathematics*, tells us that pictures or graphs were used to demonstrate a variable as early as the mid-1300's. Although coordinate systems had been used previously, this point in time seems to be the earliest record of a "graphical representation of a variable quantity."

Examples

Example 1. Find the amplitude and period of the function $y = 3 \sin 2\theta$ and sketch its graph.

Solution 1.
Step 1. For $y = 3 \sin 2\theta$ we have $A = 3$ and $B = 2$, so

$$\text{Amplitude} = A = 3 \text{ and Period} = \frac{2\pi}{B} = \frac{2\pi}{2} = \pi.$$

Step 2. On $[0, \pi]$ draw a compressed cycle of the sine function with y-values oscillating between 3 and -3. Extend the graph by duplicating this curve on successive intervals of length π as in Figure 5.17.

Step 3. If you have access to a graphics calculator, use it to graph $y = 3 \sin 2\theta$. Be sure to specify an appropriate coordinate grid (or viewing window). On the TI-82® an appropriate viewing window is $\text{Xmin} = -\pi$, $\text{Xmax} = 2\pi$, $\text{Ymin} = -3$ and $\text{Ymax} = 3$. Compare the graph in Figure 5.17 with the one generated by the calculator.

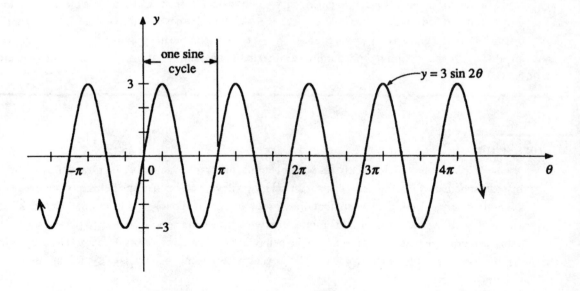

Figure 5.17 Graph of $y = 3 \sin 2\theta$

Example 2. Find the amplitude and period of the function $y = 4 \cos \dfrac{\theta}{3}$ and sketch the graph.

Solution 2.

Step 1. For $y = 4 \cos \dfrac{\theta}{3}$ we have $A = 4$ and $B = \dfrac{1}{3}$, so

$$\text{Amplitude} = A = 4 \text{ and Period} = \frac{2\pi}{B} = \frac{2\pi}{\frac{1}{3}} = 6\pi.$$

Step 2. On $[0, 6\pi]$ draw an expanded cycle of the cosine function with y-values oscillating between 4 and -4. Extend the graph by duplicating this curve on successive intervals of length 6π as in Figure 5.18.

Step 3. If you have access to a graphics calculator, use it to graph $y = 4 \cos \dfrac{\theta}{3}$. Be sure to specify a coordinate grid (or viewing window) that will show at least one cycle of the graph. Compare the graph in Figure 5.18 with the one generated by the calculator. You may need to adjust the viewing window to make the graphs look similar.

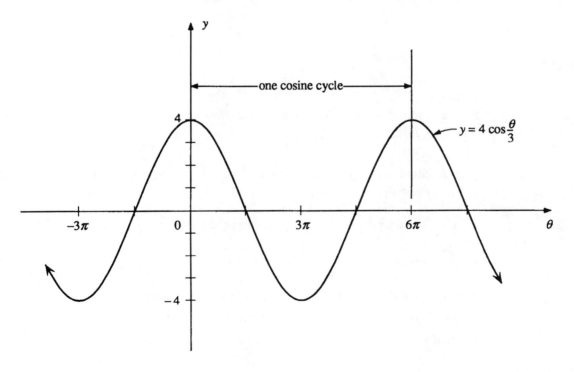

Figure 5.18 Graph of $y = 4 \cos \dfrac{\theta}{3}$

Example 3. Find the amplitude and the period of the function shown in Figure 5.19. Find the equation for the function.

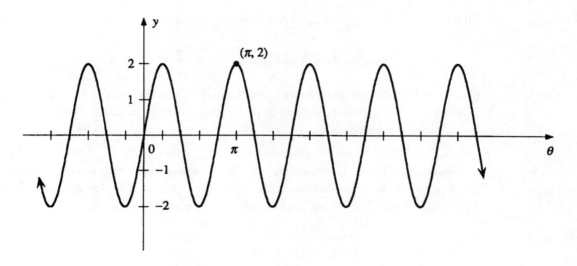

Figure 5.19

Solution 3.

Step 1. From its general shape we see that this is the graph of a function of the form $y = A \sin B\theta$ or $y = A \cos B\theta$. Since $y = 0$ when $\theta = 0$, it is the graph of $y = A \sin B\theta$.

Step 2. Since the graph oscillates two units above and below the horizontal axis, the amplitude is 2 and $A = 2$.

Step 3. The interval from 0 to $\dfrac{4\pi}{5}$ contains exactly one cycle of the graph, so the period is $\dfrac{4\pi}{5}$. This means that $\dfrac{2\pi}{B} = \dfrac{4\pi}{5}$ and $B = \dfrac{5}{2}$.

Step 4. Since $A = 2$ and $B = \dfrac{5}{2}$, the equation for the function is $y = 2 \sin \dfrac{5}{2}\theta$.

Step 5. If you have access to a graphics calculator, check the conclusion that this is the graph of $y = 2 \sin \dfrac{5}{2}\theta$ by using the calculator to graph this function and comparing the graph on the calculator with the graph in Figure 5.19. You will need to specify an appropriate viewing window to obtain graphs that can be readily compared. Use the TRACE feature to verify that the point $(\pi, 2)$ is on the graph generated by the calculator.

Example 4. Find the amplitude and the period of the function shown in Figure 5.20. Find the equation for the function.

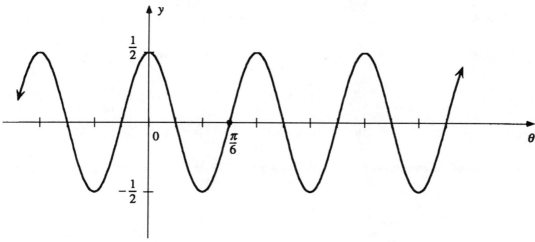

Figure 5.20

Solution 4.

Step 1. From its general shape we see that this is the graph of a function of the form $y = A \sin B\theta$ or $y = A \cos B\theta$. Since the graph has a peak at $\theta = 0$, it is the graph of $y = A \cos B\theta$.

Step 2. The graph oscillates $\dfrac{1}{2}$ unit above and below the horizontal axis, therefore the amplitude is $\dfrac{1}{2}$ and $A = \dfrac{1}{2}$.

Step 3. Since the interval from 0 to $\dfrac{4\pi}{18} = \dfrac{2\pi}{9}$ contains exactly one cycle of the graph, the period is $\dfrac{2\pi}{9}$. This means that $\dfrac{2\pi}{B} = \dfrac{2\pi}{9}$ and $B = 9$.

Step 4. Since $A = \dfrac{1}{2}$ and $B = 9$, the equation for the function is $y = \dfrac{1}{2} \cos 9\theta$.

Step 5. If you have access to a graphics calculator, check the conclusion that this is the graph of $y = \dfrac{1}{2} \cos 9\theta$ by using the calculator to graph this function and comparing the graph on the calculator with the graph in Figure 5.20. You will need to specify an appropriate viewing window to obtain graphs that can be readily compared. Use the TRACE feature to verify that the point $\left(\dfrac{\pi}{6}, 0 \right)$ is on the graph generated by the calculator.

Practice Problems

In **Problems 1 - 12**, find the amplitude and period of the functions and sketch their graphs.

1. $y = 5 \cos \theta$

2. $y = \dfrac{1}{3} \sin \theta$

3. $y = \cos \dfrac{1}{4} \theta$

4. $y = \sin 2\theta$

5. $y = 2 \cos 3\theta$

6. $y = 2 \sin \dfrac{1}{3} \theta$

7. $y = \dfrac{3}{2} \cos 2\theta$

8. $y = \dfrac{7}{3} \sin \dfrac{1}{2} \theta$

9. $y = 3 \cos \pi\theta$

10. $y = 7 \sin 2\pi\theta$

11. $y = \dfrac{5}{4} \cos \dfrac{\pi}{2} \theta$

12. $y = \dfrac{3}{2} \sin \dfrac{2\pi}{3} \theta$

13. Find the amplitude and period of the function graphed in Figure 5.21. Find the equation for the function.

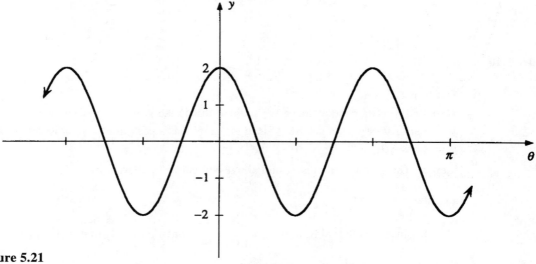

Figure 5.21

14. Find the amplitude and period of the function graphed in Figure 5.22. Find the equation for the function.

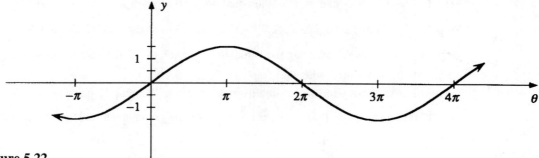

Figure 5.22

15. Find the equation for the function graphed in Figure 5.23.

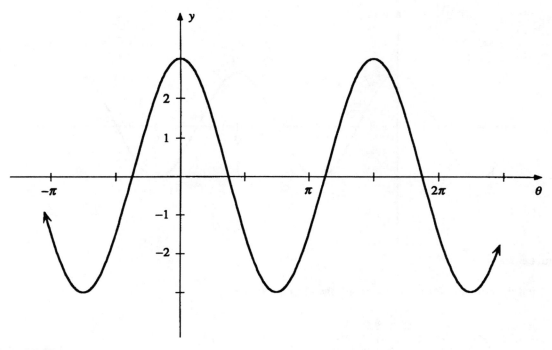

Figure 5.23

16. Find the equation for the function graphed in Figure 5.24.

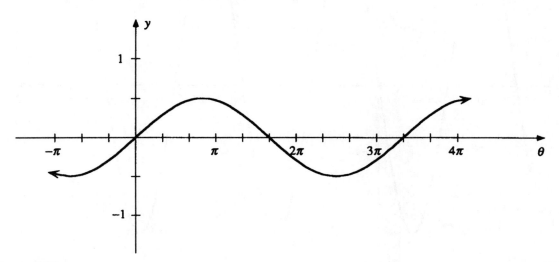

Figure 5.24

158

17. Find the equation for the function graphed in Figure 5.25.

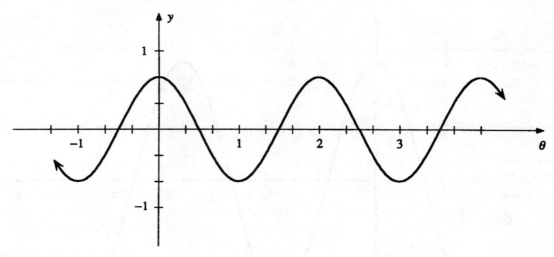

Figure 5.25

18. Find the equation for the function graphed in Figure 5.26.

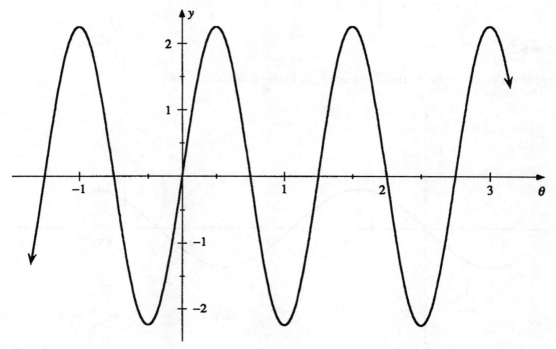

Figure 5.26

Objective 5.4

Sketch the graphs of the tangent, cotangent, cosecant and secant functions and state their periods.

Discussion

In Objective 5.2 we generated the graphs of the sine and cosine by using their values for special angles and their periodicity. In this Discussion we develop the graph of the tangent and cotangent by similar means and then obtain the graphs of the cosecant and secant by using the reciprocal relationships between these functions and the sine and cosine.

All of the trigonometric functions repeat their values every 2π units. The sine and cosine do not repeat their values over any shorter interval, so these functions have period 2π. But the tangent and cotangent are different. They have period π.

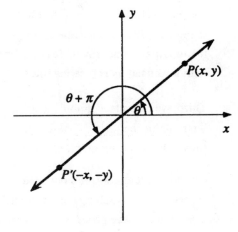

Figure 5.27

To see that the tangent and cotangent have period π, consider angles θ and $\theta + \pi$ in standard position as shown in Figure 5.27. If $P(x, y)$ is a point on the terminal side of θ, then the point $P'(-x, -y)$ lies on the terminal side of $\theta + \pi$. Thus,

$$\tan(\theta + \pi) = \frac{-y}{-x} = \frac{y}{x} = \tan\theta$$

and

$$\cot(\theta + \pi) = \frac{-x}{-y} = \frac{x}{y} = \cot\theta$$

so

(2)

$$\boxed{\begin{aligned} \tan\theta &= \tan(\theta + \pi) \\ \cot\theta &= \cot(\theta + \pi). \end{aligned}}$$

To graph the tangent function manually, we first make a table of values for angles in an interval of length π. Since the tangent is undefined for $\theta = -\dfrac{\pi}{2}$ and $\theta = \dfrac{\pi}{2}$, the interval $-\dfrac{\pi}{2} < \theta < \dfrac{\pi}{2}$ is convenient for this purpose. Table 5.7 lists values of $\tan \theta$ for angles in this interval. These values were found using a calculator, but the values for special angles can be found directly.

θ	$-\dfrac{\pi}{2}$	$-\dfrac{5\pi}{12}$	$-\dfrac{\pi}{3}$	$-\dfrac{\pi}{4}$	$-\dfrac{\pi}{6}$	$-\dfrac{\pi}{12}$	0	$\dfrac{\pi}{12}$	$\dfrac{\pi}{6}$	$\dfrac{\pi}{4}$	$\dfrac{\pi}{3}$	$\dfrac{5\pi}{12}$	$\dfrac{\pi}{2}$
$\tan \theta$	—	−3.73	−1.73	−1	−0.58	−0.27	0	0.27	0.58	1	1.73	3.73	—

Table 5.7

Second, we plot the points from Table 5.7 and draw a smooth curve through them. From plotting these points (and considering the definition of the tangent function) we see that as θ increases toward $\dfrac{\pi}{2}$ the values of the tangent function become larger and larger. The graph gets closer and closer to the vertical line $\theta = \dfrac{\pi}{2}$ but never touches it. As θ decreases toward $-\dfrac{\pi}{2}$ the values of the tangent function become larger and larger negative numbers and the graph gets closer and closer to the vertical line $\theta = -\dfrac{\pi}{2}$. The lines $\theta = \dfrac{\pi}{2}$ and $\theta = -\dfrac{\pi}{2}$ are **vertical asymptotes**. They are drawn in Figure 5.28 as dashed lines. The graph of the tangent has a vertical asymptote at every point where the function is undefined. **It is easier to draw the graph if these vertical asymptotes are drawn first and used as reference lines.**

Third, we extend the graph outside the basic interval $-\dfrac{\pi}{2} < \theta < \dfrac{\pi}{2}$ by duplicating it on successive intervals of length π. These intervals lie between successive vertical asymptotes. Figure 5.28 shows the resulting graph of the tangent function.

Graphics calculators are a powerful tool for studying the graphs of the trigonometric functions. The procedures for graphing are different for various makes and models of graphics calculators. A brief description of the procedure for graphing the tangent function of the Texas Instruments TI-82® follows. Consult the operator's manual to learn the procedure for graphing functions on the particular maching you are using or for a more detailed discussion of graphing with the Texas Instruments TI-82®.

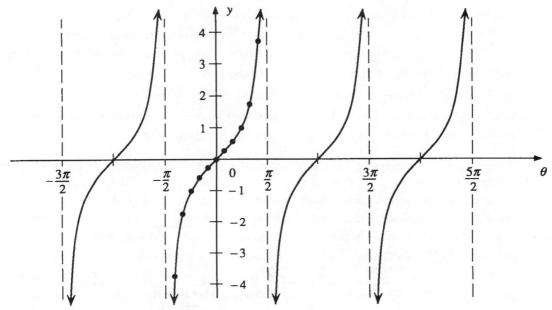

Figure 5.28 Graph of $y = \tan\theta$

To graph $y = \tan\theta$ on the Texas Instruments TI-82®, first access the mode menu by pressing MODE . Select radians as the unit of angular measure by scrolling to **Radian** and pressing ENTER . If the machine is set to connect the points of the graph, it will connect points across the vertical asymptotes, resulting in an incorrect and misleading graph. To avoid this difficulty, set the machine to **not** connect points on the graph by scrolling to **Dot** on the MODE screen and pressing ENTER .

Next, define the coordinate grid on which the graph will be displayed by specifying the largest and smallest values of x and y to be displayed. To establish $\dfrac{-3\pi}{2}$ to $\dfrac{3\pi}{2}$ as the interval over which the functions will be graphed, press WINDOW to display the WINDOW variables edit screen. Move the cursor to the line to define **Xmin**, press (−) , 3 , 2nd , [π], ÷ , 2 and ENTER . The cursor will automatically move to the line to define **Xmax**. Press 3 , 2nd , [π], ÷ , 2 and ENTER . To establish the interval −3 to 3 as the range of y values that will be displayed, move the cursor to the **Ymin** line, press (−) , 3 and ENTER . The cursor will automatically move to the **Ymax** line. Press 3 and ENTER . Next, on the Y = edit screen define the functions to be graphed. Press Y= to access this screen. Clear all functions previously entered on this screen by scrolling to each function displayed and pressing CLEAR . Position the cursor at **Y1** and enter the tangent function by pressing TAN followed by X,T,Θ . Finally, press GRAPH to generate the graph of $y = \tan\theta$. To see that the section of this graph between the vertical asymptotes at $\dfrac{-\pi}{2}$ and $\dfrac{\pi}{2}$ is repeated on successive intervals of length π, press TRACE . Then use the left and right cursor move keys, ◄ and ► , to pan across the graph.

Figure 5.29 shows the graph of the cotangent function. The graph of the cotangent also has a vertical asymptote at every point where the function is undefined. To obtain this graph manually, first draw the vertical asymptotes as dashed lines. Second, make a table of values of the cotangent for angles in the interval from 0 to π (between the first two asymptotes). Because of the reciprocal relationship between the tangent and cotangent, for θ between 0 and $\frac{\pi}{2}$ these values are the reciprocals of values for the tangent in Table 5.7. Third, graph the function on this interval. Finally, extend the graph outside this interval by duplicating it between successive asymptotes.

To graph $y = \cot \theta$ on the TI-82®, make a small change in the procedure for graphing the tangent function just given. On the Y = edit screen, replace the tangent function with the cotangent. As is the case with most scientific calculators, there is not a cotangent key on the TI-82®, so we must specify the cotangent as the reciprocal of the tangent. Press $\boxed{\text{Y=}}$ to access the Y = edit screen. Clear all functions previously entered on this screen by scrolling to each function displayed and pressing $\boxed{\text{CLEAR}}$. Position the cursor at **Y1** and enter the cotangent function as the reciprocal of the tangent by pressing $\boxed{1}$, $\boxed{\div}$, $\boxed{\text{TAN}}$ and $\boxed{\text{X,T,}\Theta}$. Finally, press $\boxed{\text{GRAPH}}$ to generate the graph of $y = \cot \theta$. To see that the section of this graph between the vertical asymptotes at 0 and π is repeated on successive intervals of length π, press $\boxed{\text{TRACE}}$. Then use the left and right cursor move keys, $\boxed{\blacktriangleleft}$ and $\boxed{\blacktriangleright}$, to pan across the graph. Consult the operator's manual to learn the procedure for graphing functions on the particular machine you are using or for a more detailed discussion of graphing with the Texas Instruments TI-82®.

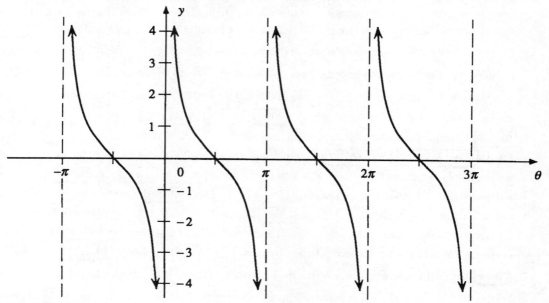

Figure 5.29 Graph of $y = \cot \theta$

Since the tangent and cotangent repeat their values over intervals of length π, they are **periodic functions**. Because they do not repeat their values over intervals of length less than π, they have period π. They are not assigned an amplitude.

The graph of $y = \csc \theta$ is shown in Figure 5.30. The relationship

(3)

$$\csc \theta = \frac{1}{\sin \theta}$$

is used to obtain the graph of $y = \csc \theta$ from the graph of $y = \sin \theta$. In order to emphasize this relationship, the graph of $y = \sin \theta$ is shown as a dashed curve in Figure 5.30.

Equation (3) shows that since the sine function is periodic with period 2π (so that it repeats itself on intervals of length 2π), the cosecant has the same property. We can obtain the entire graph of the cosecant from the portion on the interval from 0 to 2π by duplicating it on successive intervals of length 2π.

As with the tangent and cotangent, the cosecant has vertical asymptotes at the points where it is not defined. These are the points $\theta = 0, \pm \pi, \pm 2\pi$, etc. Notice from equation (3) that these are the points where the sine is zero. The vertical asymptotes for the cosecant are drawn as dashed lines in Figure 5.30.

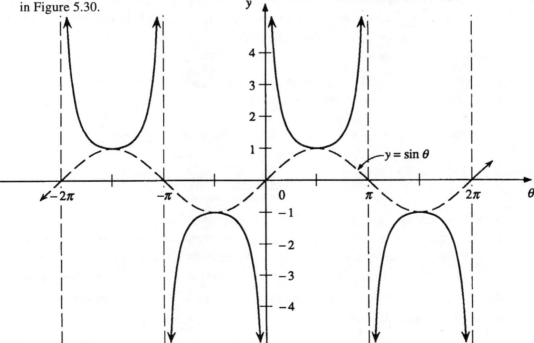

Figure 5.30 Graph of $y = \csc \theta$

To draw the graph of $y = \csc\theta$ manually, first draw its vertical asymptotes as dashed lines at the points $\theta = 0$, $\pm\pi$, $\pm 2\pi$, etc. Second, use equation (3) to find values of $\csc\theta$ for familiar values of θ between 0 and 2π. For example,

$$\csc\frac{\pi}{6} = \frac{1}{\sin{\pi}/{6}} = \frac{1}{1/2} = 2.$$

Third, plot these points. On the interval from 0 to π draw a smooth curve which passes through these points and has the lines $\theta = 0$ and $\theta = \pi$ as vertical asymptotes. Then, draw a similar curve on the interval from π to 2π. Finally, duplicate this graph on successive intervals of length 2π.

To graph $y = \csc\theta$ on the TI-82®, adapt the procedure given to graph $y = \tan\theta$ from earlier in this discussion. After defining a suitable viewing window, enter the cosecant function on the Y = edit screen. As is the case with most scientific calculators, there is not a cosecant key on the TI-82®, so we must specify the cosecant as the reciprocal of the sine. Press [Y=] to access the Y = edit screen. Clear all functions previously entered on this screen by scrolling to each function displayed and pressing [CLEAR]. Position the cursor at **Y1** and enter the cosecant function as the reciprocal of the sine by pressing [1], [÷], [SIN] and [X,T,Θ]. Finally, press [GRAPH] to generate the graph of $y = \csc\theta$. To see that the section of this graph between the vertical asymptotes at 0 and π is repeated on successive intervals of length π, press [TRACE]. Then use the left and right cursor move keys, [◄] and [►], to pan across the graph. Consult the operator's manual to learn the procedure for graphing functions on the particular machine you are using or for a more detailed discussion of graphing with the Texas Instruments TI-82®.

Notice that the values of the cosecant function are all greater than or equal to 1 or less than or equal to -1. The cosecant is not assigned an amplitude.

The graph of $y = \cos\theta$ is shown in Figure 5.31. The relationship

(**4**)

$$\sec\theta = \frac{1}{\cos\theta}$$

is used to obtain the graph of $y = \sec\theta$ from the graph of $y = \cos\theta$. Equation (4) shows that since the cosine function is periodic with period 2π, the secant has the same property. Thus, we can obtain the entire graph from the part on an interval of length 2π by duplicating this part on successive intervals. We also see from equation (4) that the secant is undefined at the points $\theta = \pm\dfrac{\pi}{2}$, $\pm\dfrac{3\pi}{2}$, $\pm\dfrac{5\pi}{2}$, etc., where the cosine is 0. As with the other trigonometric functions, the secant has vertical asymptotes at the points where it is not defined.

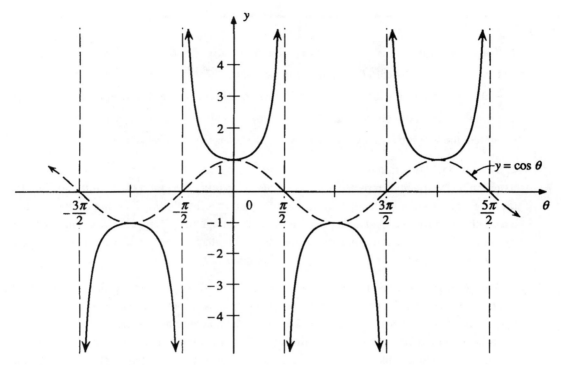

Figure 5.31 Graph of $y = \sec \theta$

To draw the graph of $y = \sec \theta$ manually, first draw vertical asymptotes at the points $\theta = \pm\dfrac{\pi}{2}$, $\pm\dfrac{3\pi}{2}$, $\pm\dfrac{5\pi}{2}$, *etc.*, where the secant is not defined. Second, use equation (4) to find values of $\sec \theta$ for familiar values of θ between 0 and 2π.

Third, plot these points and draw a smooth graph which passes through these points and has vertical asymptotes at the points $\theta = \dfrac{\pi}{2}$ and $\theta = \dfrac{3\pi}{2}$. Finally, duplicate this graph on successive intervals of length 2π.

To graph $y = \sec \theta$ on the Texas Instruments TI-82®, follow the procedure just described for graphing the cosecant, but on the Y = edit screen enter the secant function as the reciprocal of the cosine.

Notice that the graphs of the secant and cosecant are similar in the same way that the graphs of the sine and cosine are similar. The secant is not assigned an amplitude.

In this Discussion the tangent, cotangent, secant and cosecant were graphed as functions of θ in radians. It is these graphs that are used most often. To obtain their graphs as functions of θ in degrees, simply relabel the units on the horizontal axis accordingly.

Summary

1. The tangent function is periodic with period π. It is not defined for odd integer multiples of $\frac{\pi}{2}$ and its graph has vertical asymptotes at these points. Construct the graph on the interval from $-\frac{\pi}{2}$ to $\frac{\pi}{2}$ (between the most conveniently located pair of asymptotes) by plotting function values and drawing a smooth curve with vertical asymptotes at $-\frac{\pi}{2}$ and $\frac{\pi}{2}$ through the points plotted. Complete the graph by duplicating this curve on successive intervals of length π.

2. The cotangent function is periodic with period π. It is not defined for integer multiples of π and its graph has vertical asymptotes at these points. Construct the graph on the interval from 0 to π (between the most conveniently located pair of asymptotes) by plotting function values and drawing a smooth curve with vertical asymptotes at 0 and π through the points plotted. Complete the graph by duplicating this curve on successive intervals of length π. The reciprocal relation between the tangent and the cotangent is reflected in their graphs.

3. The cosecant function is periodic with period 2π. It is not defined for integer multiples of π (where the sine is 0) and its graph has vertical asymptotes at these points. The values of cosecant for familiar values of θ between 0 and 2π can be found from the values of the sine by using the reciprocal relation between these two functions. Construct the graph on the interval from 0 to 2π by plotting function values and drawing a smooth curve with vertical asymptotes at 0, π and 2π through the points plotted. Complete the graph by duplicating this curve on successive intervals of length 2π. The reciprocal relation between the sine and the cosecant is reflected in their graphs.

4. The secant function is periodic with period 2π. It is not defined for odd integer multiples of $\frac{\pi}{2}$ (where the cosine is 0) and its graph has vertical asymptotes at these points. The values of secant for familiar values of θ between 0 and 2π can be found from the values of the cosine by using the reciprocal relation between these two functions. Construct the graph on the interval from 0 to 2π by plotting function values and drawing a smooth curve with vertical asymptotes at $\frac{\pi}{2}$ and $\frac{3\pi}{2}$ through the points plotted. Complete the graph by duplicating this curve on successive intervals of length 2π. The reciprocal relation between the cosine and the secant is reflected in their graphs. The graphs of the secant and cosecant are similar in the same way that the graphs of the cosine and sine are similar.

5. To generate the graphs of the tangent, cotangent, secant and cosecant functions on an advanced scientific or graphics calculator, use the graphing procedures described in the operator's manual for the machine you are using.

Practice Problems

1. What is the period of $y = \csc \theta$?

2. What is the period of $y = \tan \theta$?

3. What is the period of $y = \cos \theta$?

4. What is the period of $y = \cot \theta$?

5. What is the period of $y = \sin \theta$?

6. What is the period of $y = \sec \theta$?

7. Using the coordinate system given in Figure 5.32 sketch the graph of the function $y = \cot \theta$ by making a table of values. Use angles θ in the interval $0 \le \theta \le \pi$ which are multiples of $\dfrac{\pi}{12}$.

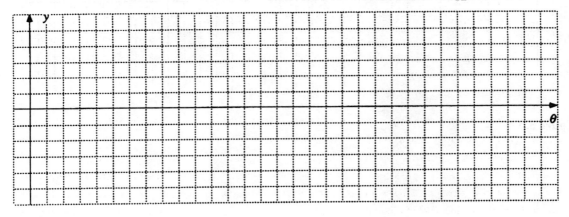

Figure 5.32

8. Using the coordinate system given in Figure 5.33 sketch the graph of the function $y = \csc \theta$ by making a table of values. Use angles θ in the interval $0 \le \theta \le 2\pi$ which are multiples of $\dfrac{\pi}{12}$.

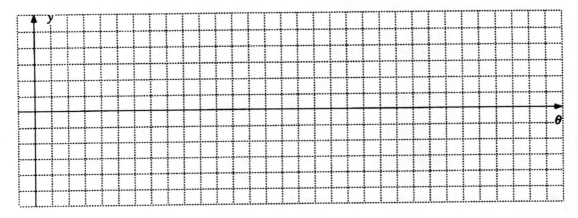

Figure 5.33

9. Using the coordinate system given in Figure 5.34 sketch the graph of the function $y = \sec \theta$ by making a table of values. Use angles θ in the interval $-\dfrac{\pi}{2} \le \theta \le \dfrac{3\pi}{2}$ which are multiples of $\dfrac{\pi}{12}$.

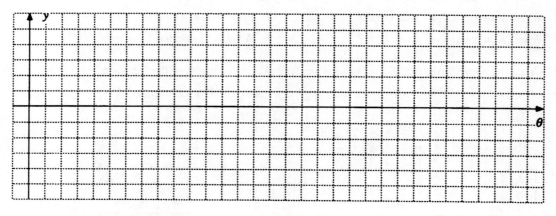

Figure 5.34

UNIT 5
Sample Examination 1

1. What is the exact value of $\sin \dfrac{\pi}{4} \cot \dfrac{\pi}{6} - \csc 60° \tan 45°$?

A) $\dfrac{\sqrt{3} - 3\sqrt{6}}{2}$　　　B) $\dfrac{3\sqrt{6} - 4\sqrt{3}}{6}$　　　C) $\dfrac{\sqrt{6} - \sqrt{3}}{2}$ D) $\dfrac{3 - \sqrt{6}}{3}$　　　E) $\dfrac{2\sqrt{3} - \sqrt{6}}{6}$

2. On the graph of $y = \sin \theta$ (θ in radians) shown in Figure 5.35, what are the coordinates of point P?

A) $\left(\dfrac{3\pi}{2}, -1\right)$

B) $\left(\dfrac{3\pi}{4}, -1\right)$

C) $\left(-1, \dfrac{3}{2}\right)$

D) $(3\pi, -1)$

E) $\left(\dfrac{3}{2}, -1\right)$

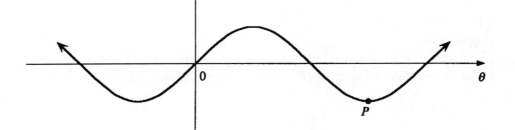

Figure 5.35

3. What is the period of the function $y = \dfrac{2}{3} \cos \dfrac{\pi}{2} \theta$ (θ in radians)?

A) $\dfrac{2}{3}$　　　B) $\dfrac{\pi}{2}$　　　C) 4　　　D) $\dfrac{1}{2}$　　　E) 1

4. The graph of a function of the form $y = A \sin B\theta$ or $y = A \cos B\theta$ (θ in radians) is shown in Figure 5.36. What is the equation for the function?

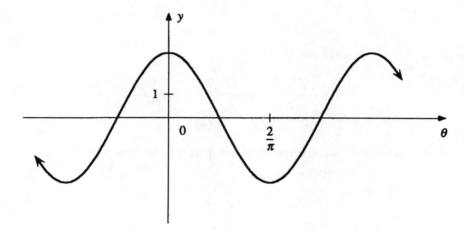

A) $y = 2 \sin \pi^2 \theta$

B) $y = 2 \cos \dfrac{1}{2}\theta$

C) $y = 2 \cos \dfrac{\pi^2}{4}\theta$

D) $y = 2 \sin 4\theta$

E) $y = 2 \cos \dfrac{\pi^2}{2}\theta$

Figure 5.36

5. Which trigonometric function (θ in radians) is graphed in Figure 5.37?

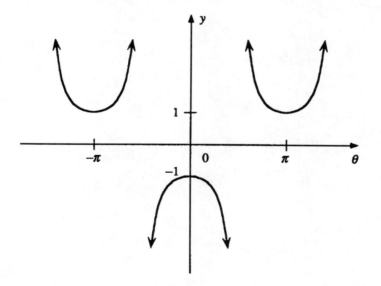

A) $y = \tan \theta$
B) $y = \cot \theta$
C) $y = \sec \theta$
D) $y = \csc \theta$
E) none of these

Figure 5.37

UNIT 5
Sample Examination 2

1. What is the exact value of $\csc \dfrac{\pi}{6} \cos \dfrac{\pi}{3} + \sin 45° \cot 30°$?

A) $\dfrac{\sqrt{6}}{2}$ **B)** $\dfrac{1+\sqrt{6}}{2}$ **C)** $\dfrac{\sqrt{3}+\sqrt{2}}{6}$ **D)** $\dfrac{2+\sqrt{6}}{2}$ **E)** $\dfrac{\sqrt{6}+3}{3}$

2. On the graph of $y = \cos \theta$ (θ in radians) shown in Figure 5.38, what are the coordinates of point Q?

A) $(2\pi, 1)$

B) $(\pi, 1)$

C) $(4\pi, 1)$

D) $(2, 1)$

E) $(3, 1)$

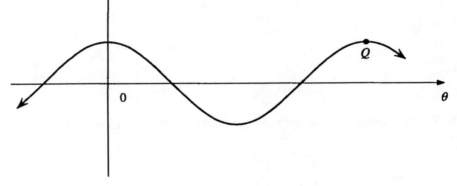

Figure 5.38

3. What is the amplitude of the function $y = 5 \sin \dfrac{1}{3}\theta$ (θ in radians)?

A) 5   **B)** $\dfrac{1}{3}$ **C)** 3   **D)** $\dfrac{1}{5}$ **E)** $\dfrac{3}{5}$

4. The graph of a function of the form $y = A \sin B\theta$ or $y = A \cos B\theta$ is shown in Figure 5.39. What is the equation for the function?

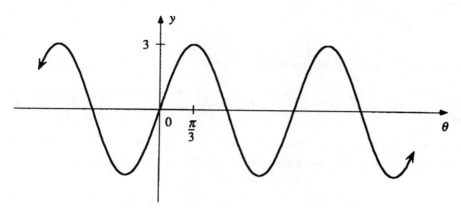

A) $y = 3 \cos \dfrac{3}{2}\theta$

B) $y = 2 \cos 3\pi\theta$

C) $y = 3 \sin \dfrac{3}{2}\theta$

D) $y = 2 \sin 3\pi\theta$

E) $y = 3 \sin \dfrac{3\pi}{2}\theta$

Figure 5.39

5. Which trigonometric function (θ in radians) is graphed in Figure 5.40?

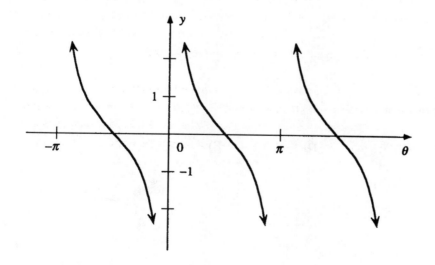

A) $y = \tan \theta$

B) $y = \cot \theta$

C) $y = \sec \theta$

D) $y = \csc \theta$

E) none of these

Figure 5.40

UNIT 6

COURSE REVIEW

Introduction

This Unit is intended to help you integrate what you have learned about numerical trigonometry. We can more readily recall, more ingeniously apply, and more efficiently expand upon information that we have internalized as a network of interrelated and mutually illuminating facts. Others can help, but finally we must develop these interactive networks for ourselves. One way to develop a rich information web is to deliberately and consciously ask yourself questions that involve connections among different concepts, terms and techniques and then read, discuss and reflect on related experiences to develop answers that connect with what you already know.

The Discussion paragraph, the Integrative Questions and the Sample Examination in this Unit are designed to help you begin developing your trigonometry network. As you work with these questions, similar integrative questions of your own may come to mind. Explore your questions. They are the very best ones for you to use to develop your trigonometry network.

UNIT 6
COURSE REVIEW

Objective 6.1

Review all of the material in Units 1 - 5.

COURSE REVIEW

Objective 6.1
Review all of the material in Units 1 - 5.

Discussion

The concepts, principles, terminology and formulas of numerical trigonometry are all interrelated. An angle is formed by rotating a ray around its initial point. Triangles are made up of sides and angles. The trigonometric functions associate numbers with angles. These numbers can be interpreted as ratios between sides of right triangles, so right triangles provide another way to think of trigonometric functions. We can also think of the trigonometric functions as describing right triangles. As soon as we think of trigonometric functions in this way, we can use them to solve right triangles. The realization that general triangles are determined by three of their parts leads us to speculate that general triangles can also be described with trigonometric functions. This speculation leads to the Laws of Sines and Cosines and to solving general triangles.

Integrative Questions

1. Devise a system of angular measure different from those discussed in Unit 1. Explain why your system is just as good as (or even better than) the system based on degrees. Why has your system (or something similar) not been incorporated into scientific calculators?

2. Two angles in standard position which have the same terminal side are called **co-terminal**. Are two co-terminal angles necessarily the same angle? How are co-terminal angles related? How are the values of the six trigonometric functions for co-terminal angles related? Why?

3. Speculate about why a triangle having a 90° angle came to be called a right triangle. (Why aren't triangles that do not have a 90° angle called wrong triangles? Do right triangles have anything in common with right whales?)

4. Explain why different points on the terminal side of an angle θ do not produce different values for the trigonometric functions (*i.e.*, for $\sin \theta$, $\cos \theta$ and $\tan \theta$).

5. The trigonometric functions are defined in terms of three quantities, x, y and r, where x and y are the coordinates of a point and $r = \sqrt{x^2 + y^2}$ is the distance between this point and the origin. How many different ratios between two of the quantities x, y and r can be formed? Show that every one of these ratios is a trigonometric functions. (Now you see why there are six trigonometric functions.)

6. In each quadrant we find that either the sine, cosine and tangent are all three positive (quadrant I) or only one of them is positive (quadrant II - sine, quadrant III - tangent, quadrant IV - cosine). Why are there no quadrants where exactly two of these functions are positive?

7. Why don't scientific calculators have keys for the secant, cosecant and cotangent?

8. (a) Why are reference angles defined with reference to the x-axis instead of the y-axis?

(b) Define the co-reference angle θ^* of an angle θ in standard position to be the smaller of the two positive angles formed by the terminal side of θ and the y-axis. How are the values of the trigonometric functions for θ related to the values of the trigonometric functions for θ^*? What does this suggest about how the trigonometric functions are named?

(c) Does it seem reasonable that there is a reciprocal relation between the tangent and cotangent ($\cot \theta = \dfrac{1}{\tan \theta}$) but not between the sine and cosine?

9. (a) In Objective 2.5 we learned to find all angles θ of less than one revolution for which the trigonometric functions have a specified value. There are usually two such angles. How are these two angles related?

(b) If the requirement that θ be less than one revolution is removed, how many such angles are there? How are they related to the angles of less than one revolution?

10. Why can't a triangle be determined from 3 angles?

11. Two triangles which have corresponding sides of the same length are called **congruent** triangles. If $\triangle ABC$ and $\triangle DEF$ are congruent, can $\triangle ABC$ necessarily be made to coincide exactly with $\triangle DEF$ by rigidly moving $\triangle ABC$ across the page (*i.e.*, translating and rotating $\triangle ABC$ in the plane)?

12. Does the Law of Sines apply to right triangles? If it does, what does it say in the case the triangle is a right triangle? Under what circumstances might it be wise to use the Law of Sines to solve a right triangle?

13. Develop the details of the argument suggested in the Discussion for Objective 3.2 to establish the Law of Sines for the case the angle A is an obtuse angle.

14. Suppose you are to solve a triangle which has two sides of length a and b and θ ($0° < \theta < 180°$) as an opposite angle. Does it make any difference whether θ is the angle opposite side a or the angle opposite side b?

15. Given two positive numbers a and b and an angle θ with $0° < \theta < 180°$, there is always a triangle which has two sides of lengths a and b and included angle θ. However, there is *not* always a triangle which has two sides of lengths a and b and θ as the angle *opposite* one of these sides. Why the difference?

16. Not every three positive numbers are the lengths of three sides of a triangle. Explain why. Is there an easy way to tell whether three positive numbers are the lengths of the sides of some triangle?

17. Does the Law of Cosines apply to right triangles? If so, what do the three different equations from the Law of Cosines say in the case the triangle is a right triangle? Under what circumstances might it be wise to use the Law of Cosines to solve a right triangle?

18. Develop the details of the argument suggested in the Discussion for Objective 4.1 to establish the Law of Cosines for the case the angle A is an obtuse angle.

19. When $\triangle ACB$ is a right triangle with right angle C, the equation $c^2 = a^2 + b^2 - 2ab \cos C$ from the Law of Cosines reduces to the Pythagorean relation $c^2 = a^2 + b^2$. Do the other two equations from the Law of Cosines reduce to simpler equations, too? If so, what are these equations?

20. Is it possible without actually calculating any additional parts to recognize that a triangle is a right triangle from
   (*i*) the lengths of its three sides;
   (*ii*) the lengths of two sides and the measure of the included angle;
   (*iii*) the lengths of two sides and the measure of the angle opposite one of them?
How?

21. In the Discussion for Objective 5.1, the values of the trigonometric functions for the special angles $\frac{\pi}{4}(= 45°)$, $\frac{\pi}{6}(= 30°)$ and $\frac{\pi}{3}(= 60°)$ are calculated by using trigonometric ratios to describe the special relationships among sides and angles of right triangles that include one of these acute angles. Evaluate the trigonometric functions for these special angles from the definitions of the trigonometric functions. (You will need to use the special properties of right triangles that have one of these angles as an acute angle.)

22. In the Discussion for Objective 5.4, the reciprocal relationships between the sine and cosecant and between the cosine and secant are used to construct the graphs of the cosecant and secant from the graphs of the sine and cosine.
   (**a**) Use the reciprocal relationship between the tangent and cotangent to construct the graph of the cotangent from the graph of the tangent.
   (**b**) Compare your construction of the graph of the cotangent function from (a) with the construction of this graph in the Discussion for Objective 5.4.

Sample Examination

The following problems sample material from Units 1 through 5 and provide review and practice with the basic concepts and techniques of numerical trigonometry.

1. What is the radian measure of a 972° angle expressed to the nearest hundredth of a radian?

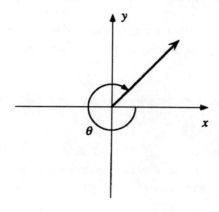

2. What is the (approximate) degree measure of the angle shown in Figure 6.1?

Figure 6.1

3. The point $(-2, 3)$ is on the terminal side of angle θ in standard position. What is the value of $3 \sin\theta + 7 \sec\theta$ rounded to two decimal places?

4. Cos $\theta = -\dfrac{5}{6}$ and $\tan\theta$ is positive. What is the value of $4 \csc\theta - \cot\theta$ rounded to three decimal places?

5. What is the value of $7 \csc(7.36) \cot\left(-\dfrac{3\pi}{8}\right) - 2 \cos(2.78)$ rounded to three decimal places?

6. What is the value of $5 \csc(212°) \cot(750°) + 9 \cos(-438°)$ rounded to four decimal places?

7. Angles S and T are between 0° and 360°, $\sin S = -0.4387$ and $\tan T = -6.8974$. What is the (approximate) value for $S + T$?

8. Angles α and β are between 0 and 2π radians, $\cos\alpha = 0.4599$ and $\sin\beta = -0.8797$. What is the (approximate) value for $\alpha + \beta$?

9. What is the (approximate) perimeter of triangle ABC, where $a = 2.32$, $A = 17°$ and $B = 90°$?

10. What is the (approximate) length of side p of triangle PQR, where $q = 12.76$, $r = 3.89$ and $Q = 90°$?

11. To the nearest tenth, what is the length of side c of triangle ABC, where $B = 57°$, $C = 101°$ and $a = 16.3$?

12. In triangle ABC, $a = 25.1$, $b = 50.4$ and $c = 33.3$. Find angle C to the nearest tenth of a degree.

13. The State Highway Commission plans to take a triangular corner plot from the square lot at the intersection of Routes 3 and 102 for a new expressway. If the plot runs 521.5 feet along Route 3 and 636.4 feet along Route 102, what is the length of the third side?

14. The top of a stairway is 2.5 meters higher than the bottom. The angle of depression from the top of the stairway to the bottom is 34°. A brass rail extends from the top of the stairway to the bottom. What is the length of the brass rail?

15. The blades on a pair of scissors are 12 centimeters long. How far apart are the tips of the scissors when the angle between the blades measures 22°?

16. A ladder 20 feet long leaned against the bottom of a second story window of a house makes an angle of 75° with the level ground. What acute angle will a 24-foot ladder leaned against the bottom of the same window make with the ground?

17. What is the exact numerical value of $\csc \dfrac{\pi}{6} \tan \dfrac{\pi}{3} - \cot 45° \sin 60°$?

18. The graph of a function of the form $y = A \sin B\theta$ or $y = A \cos B\theta$ (θ in radians) is shown in Figure 6.2. What is the equation for this function?

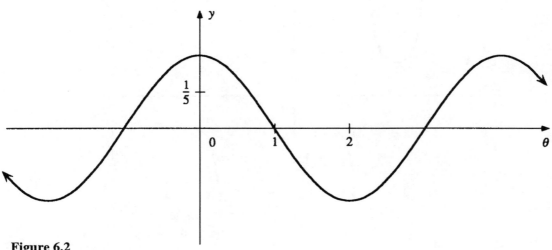

Figure 6.2

19. The graph of a function of the form $y = A \sin B\theta$ or $y = A \cos B\theta$ (θ in radians) is shown in Figure 6.3. What is the equation for this function?

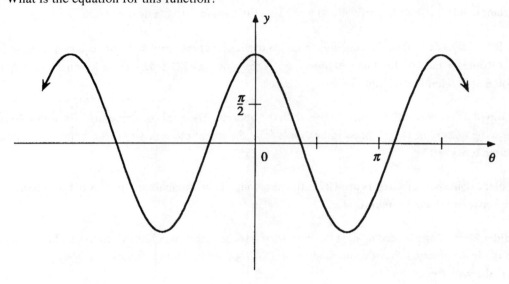

Figure 6.3

20. Which trigonometric function is illustrated by the graph shown in Figure 6.4?

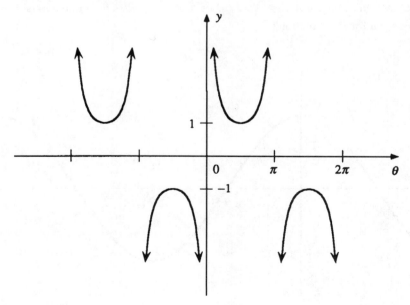

Figure 6.4

ANSWER SUPPLEMENT

NUMERICAL TRIGONOMETRY

UNIT 1
ANGLES AND TRIGONOMETRIC FUNCTIONS

Objective 1.1

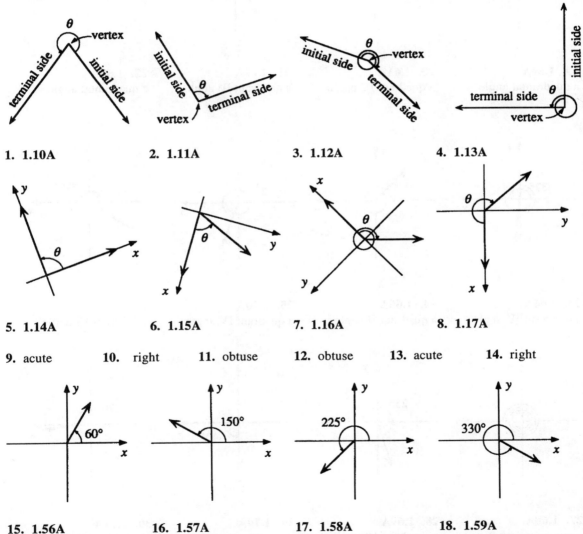

1. 1.10A   **2. 1.11A**   **3. 1.12A**   **4. 1.13A**

5. 1.14A   **6. 1.15A**   **7. 1.16A**   **8. 1.17A**

9. acute   **10.** right   **11.** obtuse   **12.** obtuse   **13.** acute   **14.** right

15. 1.56A
a quadrant I angle

16. 1.57A
a quadrant II angle

17. 1.58A
a quadrant III angle

18. 1.59A
a quadrant IV angle

184

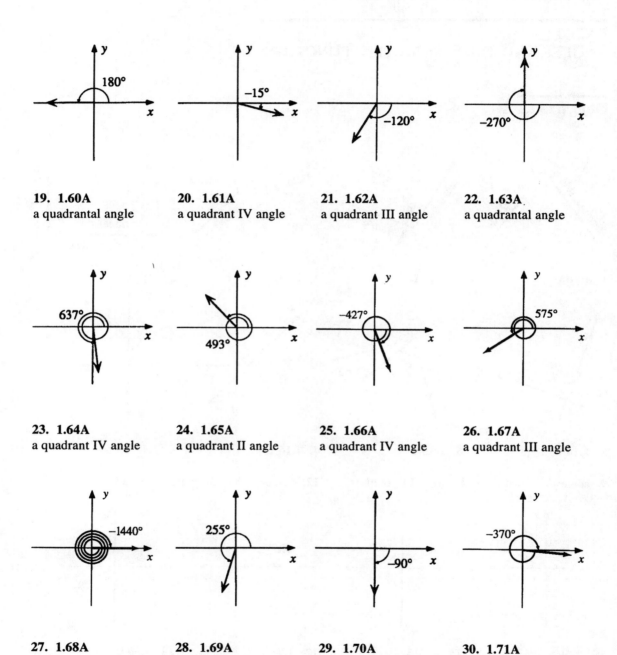

19. 1.60A
a quadrantal angle

20. 1.61A
a quadrant IV angle

21. 1.62A
a quadrant III angle

22. 1.63A
a quadrantal angle

23. 1.64A
a quadrant IV angle

24. 1.65A
a quadrant II angle

25. 1.66A
a quadrant IV angle

26. 1.67A
a quadrant III angle

27. 1.68A
a quadrantal angle

28. 1.69A
a quadrant III angle

29. 1.70A
a quadrantal angle

30. 1.71A
a quadrant IV angle

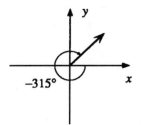

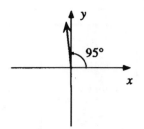

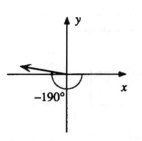

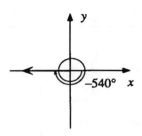

31. 1.72A
a quadrant I angle

32. 1.73A
a quadrant II angle

33. 1.74A
a quadrant II angle

34. 1.75A
a quadrantal angle

35. 405°, a quadrant I angle
36. −60°, a quadrant IV angle
37. 110°, a quadrant II angle
38. −520°, a quadrant III angle

39. 70°, a quadrant I angle
40. 630°, a quadrantal angle
41. 240°, a quadrant III angle
42. 1050°, a quadrant IV angle

43. −540°, a quadrantal angle
44. −200°, a quadrant II angle
45. (E)   **47.** (A)
46. (D)   **48.** (E)

Objective 1.2

1. $\dfrac{5\pi}{6}$ **2.** $\dfrac{5\pi}{3}$ **3.** $\dfrac{-5\pi}{4}$ **4.** 3π **5.** $\dfrac{5\pi}{12}$ **6.** $\dfrac{-\pi}{3}$ **7.** $\dfrac{-5\pi}{2}$ **8.** $\dfrac{217\pi}{180}$ **9.** $\dfrac{-17\pi}{4}$

10. 0.40
11. −3.75
12. 3.00
13. 4.31

14. 15.71
15. −1.41
16. 60°
17. 540°

18. −225°
19. 210°
20. 720°
21. 57.3°

22. −315.1°
23. −60.2°
24. 28.6°
25. 744.8°

26. −154.7°
27. 90°
28. (B)
29. (B)

Objective 1.3

	$\sin\theta$	$\cos\theta$	$\tan\theta$	$\csc\theta$	$\sec\theta$	$\cot\theta$
1.	$\dfrac{4}{5}$	$\dfrac{3}{5}$	$\dfrac{4}{3}$	$\dfrac{5}{4}$	$\dfrac{5}{3}$	$\dfrac{3}{4}$
2.	$\dfrac{12}{13}$	$\dfrac{-5}{13}$	$\dfrac{-12}{5}$	$\dfrac{13}{12}$	$\dfrac{-13}{5}$	$\dfrac{-5}{12}$
3.	$\dfrac{-7}{25}$	$\dfrac{-24}{25}$	$\dfrac{7}{24}$	$\dfrac{-25}{7}$	$\dfrac{-25}{24}$	$\dfrac{24}{7}$

	$\sin\theta$	$\cos\theta$	$\tan\theta$	$\csc\theta$	$\sec\theta$	$\cot\theta$
4.	$\dfrac{-1}{2}$	$\dfrac{\sqrt{3}}{2}$	$\dfrac{-1}{\sqrt{3}}$	-2	$\dfrac{2}{\sqrt{3}}$	$-\sqrt{3}$
5.	0	-1	0	undefined	-1	undefined
6.	$\dfrac{1}{5\sqrt{2}}$	$\dfrac{7}{5\sqrt{2}}$	$\dfrac{1}{7}$	$5\sqrt{2}$	$\dfrac{5\sqrt{2}}{7}$	7
7.	$\dfrac{7}{\sqrt{149}}$	$\dfrac{-10}{\sqrt{149}}$	$\dfrac{-7}{10}$	$\dfrac{\sqrt{149}}{7}$	$\dfrac{-\sqrt{149}}{10}$	$\dfrac{-10}{7}$
8.	-1	0	undefined	-1	undefined	0
9.	$\dfrac{3}{\sqrt{13}}$	$\dfrac{2}{\sqrt{13}}$	$\dfrac{3}{2}$	$\dfrac{\sqrt{13}}{3}$	$\dfrac{\sqrt{13}}{2}$	$\dfrac{2}{3}$
10.	$\dfrac{-3}{5}$	$\dfrac{-4}{5}$	$\dfrac{3}{4}$	$\dfrac{-5}{3}$	$\dfrac{-5}{4}$	$\dfrac{4}{3}$
11.	$\dfrac{9}{\sqrt{97}}$	$\dfrac{-4}{\sqrt{97}}$	$\dfrac{-9}{4}$	$\dfrac{\sqrt{97}}{9}$	$\dfrac{-\sqrt{97}}{4}$	$\dfrac{-4}{9}$
12.	$\dfrac{\sqrt{7}}{4}$	$\dfrac{-3}{4}$	$\dfrac{-\sqrt{7}}{3}$	$\dfrac{4}{\sqrt{7}}$	$\dfrac{-4}{3}$	$\dfrac{-3}{\sqrt{7}}$
13.	$\dfrac{\sqrt{3}}{2}$	$\dfrac{1}{2}$	$\sqrt{3}$	$\dfrac{2}{\sqrt{3}}$	2	$\dfrac{1}{\sqrt{3}}$
14.	$\dfrac{\sqrt{5}}{\sqrt{6}}$	$\dfrac{1}{\sqrt{6}}$	$\sqrt{5}$	$\dfrac{\sqrt{6}}{\sqrt{5}}$	$\sqrt{6}$	$\dfrac{1}{\sqrt{5}}$
15.	$\dfrac{-3}{\sqrt{11}}$	$\dfrac{\sqrt{2}}{\sqrt{11}}$	$\dfrac{-3}{\sqrt{2}}$	$\dfrac{-\sqrt{11}}{3}$	$\dfrac{\sqrt{11}}{\sqrt{2}}$	$\dfrac{-\sqrt{2}}{3}$

16. (A) **17. (D)**

Objective 1.4

1. I or IV **3.** I or II **5.** II **7.** II
2. III or IV **4.** II or III **6.** IV **8.** III

	$\sin\theta$	$\cos\theta$	$\tan\theta$	$\csc\theta$	$\sec\theta$	$\cot\theta$
9.	$\dfrac{1}{2}$	$\dfrac{-\sqrt{3}}{2}$	$\dfrac{-1}{\sqrt{3}}$	2	$\dfrac{-2}{\sqrt{3}}$	$-\sqrt{3}$
10.	$\dfrac{-\sqrt{3}}{2}$	$\dfrac{1}{2}$	$-\sqrt{3}$	$\dfrac{-2}{\sqrt{3}}$	2	$\dfrac{-1}{\sqrt{3}}$
11.	$\dfrac{-1}{\sqrt{10}}$	$\dfrac{-3}{\sqrt{10}}$	$\dfrac{1}{3}$	$-\sqrt{10}$	$\dfrac{-\sqrt{10}}{3}$	3
12.	$\dfrac{1}{\sqrt{37}}$	$\dfrac{6}{\sqrt{37}}$	$\dfrac{1}{6}$	$\sqrt{37}$	$\dfrac{\sqrt{37}}{6}$	6
13.	$\dfrac{-12}{13}$	$\dfrac{-5}{13}$	$\dfrac{12}{5}$	$\dfrac{-13}{12}$	$\dfrac{-13}{5}$	$\dfrac{5}{12}$
14.	$\dfrac{-2}{7}$	$\dfrac{3\sqrt{5}}{7}$	$\dfrac{-2}{3\sqrt{5}}$	$\dfrac{-7}{2}$	$\dfrac{7}{3\sqrt{5}}$	$\dfrac{-3\sqrt{5}}{2}$
15.	$\dfrac{15}{17}$	$\dfrac{-8}{17}$	$\dfrac{-15}{8}$	$\dfrac{17}{15}$	$\dfrac{-17}{8}$	$\dfrac{-8}{15}$
16.	$\dfrac{-12}{13}$	$\dfrac{5}{13}$	$\dfrac{-12}{5}$	$\dfrac{-13}{12}$	$\dfrac{13}{5}$	$\dfrac{-5}{12}$
17.	$\dfrac{-8}{17}$	$\dfrac{-15}{17}$	$\dfrac{8}{15}$	$\dfrac{-17}{8}$	$\dfrac{-17}{15}$	$\dfrac{15}{8}$
18.	$\dfrac{\sqrt{21}}{5}$	$\dfrac{-2}{5}$	$\dfrac{-\sqrt{21}}{2}$	$\dfrac{5}{\sqrt{21}}$	$\dfrac{-5}{2}$	$\dfrac{-2}{\sqrt{21}}$
19.	$\dfrac{-5}{12}$	$\dfrac{-\sqrt{119}}{12}$	$\dfrac{5}{\sqrt{119}}$	$\dfrac{12}{-5}$	$\dfrac{12}{-\sqrt{119}}$	$\dfrac{\sqrt{119}}{5}$
20.	$\dfrac{2}{\sqrt{5}}$	$\dfrac{1}{\sqrt{5}}$	2	$\dfrac{\sqrt{5}}{2}$	$\sqrt{5}$	$\dfrac{1}{2}$
21.	$\dfrac{-3\sqrt{5}}{7}$	$\dfrac{-2}{7}$	$\dfrac{3\sqrt{5}}{2}$	$\dfrac{-7}{3\sqrt{5}}$	$\dfrac{-7}{2}$	$\dfrac{2}{3\sqrt{5}}$
22.	$\dfrac{2}{5}$	$\dfrac{\sqrt{21}}{5}$	$\dfrac{2}{\sqrt{21}}$	$\dfrac{5}{2}$	$\dfrac{5}{\sqrt{21}}$	$\dfrac{\sqrt{21}}{2}$

23. (E)   **24. (B)**

Sample Examination 1

1. This positive angle is formed by $1\frac{7}{8}$ revolutions counterclockwise from the positive x-axis. Therefore, the degree measure of this angle is $\left(1\frac{7}{8}\right)360° = \left(\frac{15}{8}\right)360° = 675°$.

2. -0.55 radians $= -0.55\left(\frac{180}{\pi}\right) = -31.51$ (to the nearest hundredth).

3. Since the point $(9, 6)$ is on the terminal side of θ, choose $x = 9$, $y = 6$ and $r = \sqrt{81+36} = \sqrt{117}$. Then, $\tan\theta = \frac{y}{x} = \frac{6}{9}$, $\csc\theta = \frac{r}{y} = \frac{\sqrt{117}}{6}$ and, rounded to one decimal place,

$$8\tan\theta + 5\csc\theta = 8\left(\frac{6}{9}\right) + 5\left(\frac{\sqrt{117}}{6}\right) = 14.3.$$

4. The point $(-5, -8)$ is on the terminal side of θ (it is in quadrant III and $\tan\theta = \frac{y}{x} = \frac{8}{5}$). We have $x = -5$, $y = -8$, and $r = \sqrt{25+64} = \sqrt{89}$. Thus, $\sin\theta = \frac{y}{r} = \frac{-8}{\sqrt{89}}$, $\sec\theta = \frac{r}{x} = \frac{\sqrt{89}}{-5}$ and, rounded to three decimal places,

$$7\sin\theta + 8\sec\theta = 7\left(\frac{-8}{\sqrt{89}}\right) + 8\left(\frac{\sqrt{89}}{-5}\right) = -21.030.$$

5. Since $\cot\theta > 0$ and $\sin\theta < 0$, θ is a quadrant III angle. Choose (x, y) on the terminal side of θ so $x < 0$, $y < 0$ and $\cot\theta = \frac{x}{y} = \frac{11}{2}$. The easiest choice is $x = -11$, $y = -2$, $r = \sqrt{121+4}$ $= \sqrt{125}$. Then, $\cos\theta = \frac{x}{r} = \frac{-11}{5\sqrt{5}}$, $\csc\theta = \frac{r}{y} = \frac{5\sqrt{5}}{-2}$ and, rounded to three decimal places,

$$5\cos\theta + 7\csc\theta = 5\left(\frac{-11}{5\sqrt{5}}\right) + 7\left(\frac{5\sqrt{5}}{-2}\right) = -44.051.$$

Sample Examination 2

1. The negative angle $-525°$ is formed by $\frac{525}{360} = 1\frac{11}{24}$ revolutions clockwise from the positive x-axis. Figure 1.52A shows this angle.

2. $-792° = -792\left(\frac{\pi}{180}\right) = -13.8$ radians (to the nearest tenth).

3. Since the point $(6, -5)$ is on the terminal side of θ, choose $x = 6$, $y = -5$ and $r = \sqrt{36 + 25} = \sqrt{61}$.

Then, $\sin \theta = \dfrac{y}{r} = \dfrac{-5}{\sqrt{61}}$, $\sec \theta = \dfrac{r}{x} = \dfrac{\sqrt{61}}{6}$ and, rounded to one decimal place,

$$\sin \theta - 3 \sec \theta = \frac{-5}{\sqrt{61}} - 3\left(\frac{\sqrt{61}}{6}\right) = -4.5.$$

4. The point $\left(\sqrt{187}, 3\right)$ is on the terminal side of θ (it is in quadrant I, $r = \sqrt{187 + 9} = 14$ and

$\sin \theta = \dfrac{y}{r} = \dfrac{3}{14}$), and $x = \sqrt{187}$, $y = 3$ and $r = 14$. Thus, $\cos \theta = \dfrac{x}{r} = \dfrac{\sqrt{187}}{14}$, $\cot \theta = \dfrac{x}{y} = \dfrac{\sqrt{187}}{3}$
and, rounded to three decimal places,

$$3 \cos \theta + 5 \cot \theta = 3\left(\frac{\sqrt{187}}{14}\right) + 5\left(\frac{\sqrt{187}}{3}\right) = 25.722.$$

5. Since $\cos \theta > 0$ and $\sin \theta < 0$, θ is a quadrant IV angle. Choose (x, y) on the terminal side of θ so

$x > 0$, $y < 0$ and $\cos \theta = \dfrac{x}{r} = \dfrac{2}{7}$. The easiest choice is $x = 2$, $r = 7$, $y = -\sqrt{49 - 4} = -\sqrt{45}$

$= -3\sqrt{5}$. Then, $\tan \theta = \dfrac{y}{x} = \dfrac{-3\sqrt{5}}{2}$, $\csc \theta = \dfrac{r}{y} = \dfrac{7}{-3\sqrt{5}}$ and, rounded to four decimal places,

$$8 \tan \theta + 2 \csc \theta = 8\left(\frac{-3\sqrt{5}}{2}\right) + 2\left(\frac{7}{-3\sqrt{5}}\right) = -28.9198.$$

UNIT 2
EVALUATING TRIGONOMETRIC FUNCTIONS

Objective 2.1

1. 0	**7.** undefined	**13.** undefined	**19.** (B)
2. undefined	**8.** 1	**14.** 0	**20.** (D)
3. 0	**9.** undefined	**15.** undefined	
4. 6	**10.** undefined	**16.** –5	
5. 3	**11.** 0	**17.** $\dfrac{-5}{3}$	
6. undefined	**12.** –2	**18.** undefined	

190

Objective 2.2

1. 0.3420	**6.** 2.1105	**11.** 8.9252	**16.** 1.4534	**21.** (B)
2. 1	**7.** −1.4826	**12.** 5.6569	**17.** −1.8165	**22.** (C)
3. undefined	**8.** −0.5	**13.** −0.5295	**18.** −9.9591	
4. 1.5523	**9.** 0.5774	**14.** 1.9367	**19.** −12.9029	
5. 1.4142	**10.** undefined	**15.** −0.9976	**20.** 4.4693	

Objective 2.3

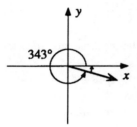

1. 2.11A
reference angle: 17°

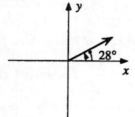

2. 2.12A
reference angle: $\frac{\pi}{8}$

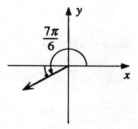

3. 2.13A
reference angle: 28°

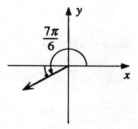

4. 2.14A
reference angle: $\frac{\pi}{6}$

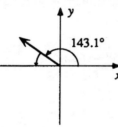

5. 2.15A
reference angle: 36.9°

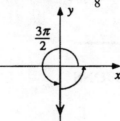

6. 2.16A
reference angle: $\frac{\pi}{2}$

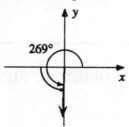

7. 2.17A
reference angle: 89°

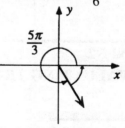

8. 2.18A
reference angle: $\frac{\pi}{3}$

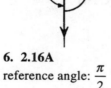

9. 2.19A
reference angle: 14°

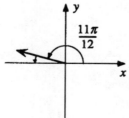

10. 2.20A
reference angle: $\frac{\pi}{12}$

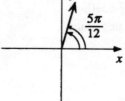

11. 2.21A
reference angle: $\frac{5\pi}{12}$

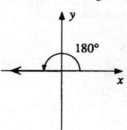

12. 2.22A
reference angle: 0°

13. (B)   **14. (B), (C)**

Objective 2.4

1. $-\tan 19°$

2. $\csc \dfrac{\pi}{8}$

3. $\cos 28°$

4. $-\sin \dfrac{\pi}{6}$

5. $-\cot 38°$

6. $\sec 36°$

7. $-\csc 19.9°$

8. $\cot \dfrac{2\pi}{7}$

9. $\sin \dfrac{5\pi}{12}$

10. $\tan 85°$

11. $-\cos \dfrac{\pi}{12}$

12. $-\sec 23°$

13. $\sin 7°$

14. $\tan \dfrac{\pi}{3}$

15. $-\sec \dfrac{6\pi}{29}$

16. $-\cos 19°$

17. $\csc \dfrac{2\pi}{7}$

18. $-\cot 32°$

19. $-\sin 1°$

20. $-\tan \dfrac{3\pi}{7}$

21. $\sec \dfrac{3\pi}{7}$

22. $-\cos \dfrac{2\pi}{15}$

23. $-\csc 73°$

24. $\cot \dfrac{5\pi}{11}$

25. (D)

26. (A), (C)

Objective 2.5

1. 30°

2. 71.93°

3. 1.45 radians

4. 0.72 radians

5. 62.24°

6. 41.76°

7. no such angles

8. no such angles

9. 3.6652 and 5.7596 radians

10. 39.7442° and 219.7442°

11. 0.7854 and 5.4978 radians

12. 159.6451° and 339.6451°

13. 0.7854 and 2.3562 radians

14. 156.9261° and 203.0739°

15. no such angles

16. no such angles

17. 90°, 246°, 270° and 426°

18. 4.7123, 5.7596, 9.9482 and 10.9955

19. 5.2622, 6.0475, 8.3938, 9.1891

20. 207°, 341°, 387°, 521°

21. (D), (E)   **22. (B), (D), (E)**

Sample Examination 1

1. Set degree mode. Rounded to three decimal places,

$$6 \cos (570°) \sec (170°) + 5 \tan (-230°) = \frac{6 \cos (570°)}{\cos (170°)} + 5 \tan (-230°) = -0.682 \, .$$

2. The reference angle for $\theta = \dfrac{3\pi}{16}$ is $\theta' = \dfrac{3\pi}{16}$.

The reference angle for $\theta = \dfrac{11\pi}{16}$ is $\theta' = \pi - \dfrac{11\pi}{16} = \dfrac{5\pi}{16}$.

The reference angle for $\theta = \dfrac{21\pi}{16}$ is $\theta' = \dfrac{21\pi}{16} - \pi = \dfrac{5\pi}{16}$.

3. $\theta = 296°$ is a quadrant IV angle so

$$\sin \theta = -\sin \theta', \qquad \csc \theta = -\csc \theta',$$
$$\cos \theta = \cos \theta', \qquad \sec \theta = \sec \theta',$$
$$\tan \theta = -\tan \theta', \qquad \cot \theta = -\cot \theta'.$$

4. Set degree mode. Since $\sin \alpha = -0.7771$, α is a quadrant III or quadrant IV angle with reference angle $\alpha' = 51°$. The possible values of α are $\alpha_1 = 231°$ and $\alpha_2 = 309°$.

Since $\cos \beta = -0.9659$, β is a quadrant II or quadrant III angle with reference angle $\beta' = 15°$. The possible values of β are $\beta_1 = 165°$ and $\beta_2 = 195°$.

The possible values of $\alpha + \beta$ are

$$\alpha_1 + \beta_1 = 396°, \qquad \alpha_1 + \beta_2 = 426°, \qquad \alpha_2 + \beta_1 = 474°, \qquad \alpha_2 + \beta_2 = 504°.$$

5. Set radian mode. Since $\sin \phi = -0.3466$, ϕ is a quadrant III or quadrant IV angle with reference angle $\phi' = 0.3539$. The possible values of ϕ are $\phi_1 = 3.4955$ and $\phi_2 = 5.9293$.

Since $\tan \psi = 0.3374$, ψ is a quadrant I or quadrant III angle with reference angle $\psi = 0.3254$. The possible values of ψ are $\psi_1 = 0.3254$ and $\psi_2 = 3.4670$.

The possible values of $\phi + \psi$, rounded to three decimal places, are

$$\phi_1 + \psi_1 = 3.821, \qquad \phi_1 + \psi_2 = 6.963, \qquad \phi_2 + \psi_1 = 6.255, \qquad \phi_2 + \psi_2 = 9.396.$$

Sample Examination 2

1. Set radian mode. Rounded to three decimal places

$$7 \tan \left(\frac{8\pi}{11}\right) \cot (-7.4) + \cos \left(\frac{29\pi}{5}\right) = \frac{7 \tan (2.28479)}{\tan (-7.4)} + \cos (18.22124) = 4.751.$$

2. $\theta = 295°$ is a quadrant IV angle with reference angle $\theta' = 65°$. For a quadrant IV angle, $\tan \theta = -\tan \theta'$. Thus, $\tan 295° = -\tan 65°$.

3. $\theta = \dfrac{11\pi}{7}$ is a quadrant IV angle so

$$\sin \theta = -\sin \theta', \qquad \csc \theta = -\csc \theta',$$
$$\cos \theta = \cos \theta', \qquad \sec \theta = \sec \theta',$$
$$\tan \theta = -\tan \theta', \qquad \cot \theta = -\cot \theta'.$$

4. Set radian mode. Since $\sin \phi = -0.1709$, ϕ is a quadrant III or quadrant IV angle with reference angle $\phi' = 0.1717$.

The possible values for ϕ are $\phi_1 = 3.3133$ and $\phi_2 = 6.1114$.

Since $\cos \psi = 0.2821$, ψ is a quadrant I or quadrant IV angle with reference angle $\psi' = 1.2848$.

The possible values for ψ are $\psi_1 = 1.2848$ and $\psi_2 = 4.9984$.

The possible values for $\phi + \psi$, rounded to three decimal places, are

$$\phi_1 + \psi_1 = 4.598, \qquad \phi_1 + \psi_2 = 8.312, \qquad \phi_2 + \psi_1 = 7.396, \qquad \phi_2 + \psi_2 = 11.110.$$

5. Set degree mode. Since $\cos R = -0.9877$, R is a quadrant II or quadrant III angle with reference angle $R' = 9°$. The possible values for R are $R_1 = 171°$ and $R_2 = 189°$.

Since $\tan S = -0.2126$, S is a quadrant II or quadrant IV angle with reference angle $S' = 12°$. The possible values for S are $S_1 = 168°$ and $S_2 = 348°$.

The possible values for $R + S$ are

$R_1 + S_1 = 339°$, $\qquad R_1 + S_2 = 519°$, $\qquad R_2 + S_1 = 357°$, $\qquad R_2 + S_2 = 537°$.

UNIT 3
RIGHT TRIANGLES AND THE LAW OF SINES

Objective 3.1

1. $c = 1.43$

2. $a = 6.81$ $\quad b = 3.37$ $\quad c = 7.60$
   $A = 63.7°$ $\quad B = 26.3°$ $\quad C = 90°$

3. $a = 2.95$

4. $a = 14.3$ $\quad b = 18.3$ $\quad c = 23.2$
   $A = 38°$ $\quad B = 52°$ $\quad C = 90°$

5. $a = 155.9$ $\quad b = 170.5$ $\quad c = 231$
   $A = 42.4°$ $\quad B = 47.6°$ $\quad C = 90°$

6. $a = 0.29$ $\quad b = 0.41$ $\quad c = 0.50$
   $A = 35°$ $\quad B = 55°$ $\quad C = 90°$

7. $a = 17.8$ $\quad b = 57.1$ $\quad c = 59.9$
   $A = 17.3°$ $\quad B = 72.7°$ $\quad C = 90°$

8. $a = 83.5$ $\quad b = 17.2$ $\quad c = 85.3$
   $A = 78.4°$ $\quad B = 11.6°$ $\quad C = 90°$

9. $r = 3.6$ $\quad s = 5.3$ $\quad t = 3.9$
   $R = 43°$ $\quad S = 90°$ $\quad T = 47°$

10. $m = 0.13$ $\quad b = 0.71$ $\quad l = 0.70$
    $M = 10.5°$ $\quad B = 90°$ $\quad L = 79.5°$

11. $n = 69$ $\quad b = 101$ $\quad c = 73.76$
    $N = 43.1°$ $\quad B = 90°$ $\quad C = 46.9°$

12. $c = 56.57$ $\quad b = 66$ $\quad s = 34$
    $C = 59°$ $\quad B = 90°$ $\quad S = 31°$

13. $r = 52.3$ $\quad e = 53.7$ $\quad g = 12.3$
    $R = 76.8°$ $\quad E = 90°$ $\quad G = 13.2°$

14. $P = 67.4°$ $\quad D = 90°$ $\quad Q = 22.6°$

15. perimeter $= 314.9$

16. perimeter $= 35.8$

17. $D = 50.6°$ $\quad G = 39.4°$

18. $a = 18.6$

19. $A = 30°$

20. $b = 20.5$

21. **(D)** $\qquad$ 22. **(A)** $\qquad$ 23. **(C)**

Objective 3.2

1. $a = 11.47$

2. $b = 20.4$

3. $b = 10.12$

4. $c = 46.02$

5. $a = 182.91$

6. $p = 77.5$ $\quad q = 134.4$ $\quad r = 72.1$
   $P = 27°$ $\quad Q = 128°$ $\quad R = 25°$

7. $k = 0.27$ $\quad m = 0.33$ $\quad n = 0.35$
   $K = 46.9°$ $\quad M = 63°$ $\quad N = 70.1°$

8. $d = 26.1$ $\quad a = 43.9$ $\quad f = 38.6$
   $D = 36.1°$ $\quad A = 83°$ $\quad F = 60.9°$

9. **(E)**

10. **(A)**

Objective 3.3

1. $a = 221$ $b = 437.7$ $c = 543$
   $A = 23°$ $B = 50.7°$ $C = 106.3°$
or
   $a = 221$ $b = 561.7$ $c = 543$
   $A = 23°$ $B = 83.3°$ $C = 73.7°$

2. no solution

3. $a = 40$ $b = 63.5$ $c = 47$
   $A = 39°$ $B = 93.3°$ $C = 47.7°$
or
   $a = 40$ $b = 9.6$ $c = 47$
   $A = 39°$ $B = 8.7°$ $C = 132.3°$

4. $a = 5.8$ $b = 4.6$ $c = 6.2$
   $A = 63°$ $B = 44.7°$ $C = 72.3°$
or
   $a = 5.8$ $b = 1.0$ $c = 6.2$
   $A = 63°$ $B = 9.3°$ $C = 107.7°$

5. no solution

6. $a = 9.2$ $b = 4.2$ $c = 7.6$
   $A = 98.6°$ $B = 26.6°$ $C = 54.8°$

7. no solution

8. $a = 4.4$ $b = 8.1$ $c = 8.3$
   $A = 31°$ $B = 72°$ $C = 77°$
or
   $a = 0.7$ $b = 8.1$ $c = 8.3$
   $A = 5°$ $B = 72°$ $C = 103°$

9. $p = 1.7$ $q = 17.3$ $r = 15.7$
   $P = 1.5°$ $Q = 164.3°$ $R = 14.2°$
or
   $p = 31.9$ $q = 17.3$ $r = 15.7$
   $P = 150.1°$ $Q = 15.7°$ $R = 14.2°$

10. $g = 62.2$ $e = 111$ $f = 57.5$
   $G = 22.9°$ $E = 136°$ $F = 21.1°$

11. no solution

12. $t = 14.7$

13. $d = 378.3$ $r = 311.9$ $t = 205.3$
   $D = 91.6°$ $R = 55.5°$ $T = 32.9°$

14. no solution

15. perimeter $= 137.9$ or 104.9

16. $C = 84.4°$ or $23.6°$

17. $b = 70$ $c = 96$ $d = 29.4$
   $B = 23.4°$ $C = 147°$ $D = 9.6°$

18. perimeter $= 79.5$ or 43.2

19. (E)

20. (E)

Sample Examination 1

1. Sketch and label the triangle.

$$\tan 5° = \frac{b}{51.13} \quad \text{so} \quad b = 51.13 \tan 5° = 4.4733$$

$$\cos 5° = \frac{51.13}{c} \quad \text{so} \quad c = \frac{51.13}{\cos 5°} = 51.3253$$

$$\text{Perimeter} = 51.13 + 4.4733 + 51.3253 = 106.9286$$

2. Sketch and label the triangle.

$$\sin A = \frac{0.52}{0.59} = 0.88136 \quad \text{so} \quad A = 61.81°$$

$$\cos B = \frac{0.52}{0.59} = 0.88136 \quad \text{so} \quad B = 28.19°$$

(Alternately, $B = 90° - A = 90° - 61.81° = 28.19°$.)

$$B - A = 28.19° - 61.81° = -33.62°$$

3. Sketch and label the triangle.

$$q = \sqrt{(9.13)^2 + (3.91)^2} = 9.9320$$

4. Sketch and label the triangle.

$$C = 180° - A - B = 180° - 89° - 17° = 74°$$

By the Law of Sines,

$$\frac{a}{\sin 89°} = \frac{800}{\sin 74°} \quad \text{and} \quad a = \frac{800 \sin 89°}{\sin 74°} = 832.1128.$$

5. Sketch and label the triangle.
By the Law of Sines,

$$\frac{\sin A}{316°} = \frac{\sin 67°}{592°} \quad \text{and} \quad \sin A = \frac{316 \sin 67°}{592°} = 0.4914.$$

Then, the possible solutions are

$$A - A_1 = 29.43° \quad \text{or} \quad A = A_2 = 180° - A_1 = 150.57°.$$

Since $A_1 + C = 96.43° < 180°$, A_1 is a solution.
Since $A_2 + C = 217.57° > 180°$, A_2 is not a solution.

Sample Examination 2

1. Sketch and label the triangle.

$$\sin 82° = \frac{a}{0.42} \quad \text{so} \quad a = 0.42 \sin 82° = 0.4159$$

$$\cos 82° = \frac{b}{0.42} \quad \text{so} \quad b = 0.42 \cos 82° = 0.0585$$

Perimeter $= 0.4159 + 0.0585 + 0.42 = 0.8944$

2. Sketch and label the triangle.

$$\tan A = \frac{12.39}{46.76} = 0.26497 \quad \text{so} \quad A = 14.84°$$

$$\tan B = \frac{46.76}{12.39} = 3.77401 \quad \text{so} \quad B = 75.16°$$

(Alternately, $B = 90° - A = 90° - 14.84° = 75.16°$.)

$$A - B = 14.84° - 75.16° = -60.32°$$

3. Sketch and label the triangle.

$$h = \sqrt{(9.03)^2 - (3.61)^2} = 8.2770$$

4. Sketch and label the triangle.

$$A = 180° - B - C = 180° - 94° - 54° = 32°$$

By the Law of Sines,

$$\frac{b}{\sin 54°} = \frac{24.6}{\sin 32°} \quad \text{so} \quad b = \frac{24.6 \sin 54°}{\sin 32°} = 37.5563.$$

5. Sketch and label the triangle.

By the Law of Sines, we must have

$$\frac{\sin B}{731.6} = \frac{\sin 118°}{683.9} \quad \text{so} \quad \sin B = \frac{731.6 \sin 118°}{683.9} = 0.94453, \text{ and } B = 70.83°.$$

Then, $B + C = 70.83° + 118° = 188.83° > 180°$. There is no solution.

UNIT 4
THE LAW OF COSINES AND APPLIED PROBLEMS

Objective 4.1

1. $a = 39$

2. $A = 111.6°$

3. no solution

4. $a = 0.8$ $b = 1.06$ $c = 1.7$
   $A = 20.74°$ $B = 28.07°$ $C = 131.19°$

5. $a = 372.82$

6. $R = 68.1°$

7. $p = 7$ $a = 24$ $s = 25$
   $P = 16.3°$ $A = 73.7°$ $S = 90°$

8. $x = 1.5$ $y = 3.6$ $z = 3.9$
   $X = 22.6°$ $Y = 67.4°$ $Z = 90°$

9. $A = 104.3°$

10. $a = 7$ $b = 7$ $c = 7$
    $A = 60°$ $B = 60°$ $C = 60°$

11. $a = 70.47$ $b = 29$ $c = 51$
    $A = 121°$ $B = 20.7°$ $C = 38.3°$

12. $a = 158.89$ $b = 179$ $c = 236$
    $A = 42.3°$ $B = 49.3°$ $C = 88.4°$

13. $a = 2.1$ $b = 1.7$ $c = 0.8$
    $A = 108.9°$ $B = 50°$ $C = 21.1°$

14. no solution

15. $a = 4.8$ $b = 5$ $c = 1.4$
    $A = 73.74°$ $B = 90°$ $C = 16.26°$

Objective 4.2

1. (a) Solve for a using the Pythagorean relation. Solve for A and C using trigonometric ratios.
   (b) $a = 33.94$ $b = 36$ $c = 12$
         $A = 70.53°$ $B = 90°$ $C = 19.47°$

2. (a) Solve for C using the Law of Sines. Solve for B using $A + B + C = 180°$. Solve for b using the Law of Sines.

$a = 15$	$b = 10.60$	$c = 6$
$A = 127°$	$B = 34.37°$	$C = 18.63°$

3. (a) Solve for C using $A + B = 90°$. Solve for b and c using trigonometric ratios.

(b)

$a = 9.5$	$b = 8.39$	$c = 4.46$
$A = 90°$	$B = 62°$	$C = 28°$

4. (a) Solve for A and B using the Law of Cosines. Solve for C using $A + B + C = 180°$.

(b)

$a = 28$	$b = 21$	$c = 35$
$A = 53.1°$	$B = 36.9°$	$C = 90°$

5. (a) Solve for B using $A + B = 90°$. Solve for b and c using trigonometric ratios.

(b)

$a = 23$	$b = 38.28$	$c = 44.66$
$A = 31°$	$B = 59°$	$C = 90°$

6. (a) Solve for B using $A + B + C = 180°$. Solve for a using the Law of Sines. Solve for b using the Law of Cosines.

(b)

$a = 14.03$	$b = 39.06$	$c = 31$
$A = 19°$	$B = 115°$	$C = 46°$

7. (a) Solve for c using the Pythagorean Theorem. Solve for A using trigonometric ratios. Solve for B using $A + B = 90°$.

(b)

$a = 17$	$b = 30$	$c = 34.48$
$A = 29.5°$	$B = 60.5°$	$C = 90°$

8. (a) Solve for c using the Law of Cosines. Solve for A using the Law of Cosines. Solve for B using $A + B + C = 180°$.

(b)

$a = 10$	$b = 11$	$c = 19.04$
$A = 23.7°$	$B = 26.3°$	$C = 130°$

9. (a) Solve for R using the Law of Sines. Solve for S using $R + S + T = 180°$. Solve for s using the Law of Cosines.

(b)

$r = 18$	$s_1 = 24.2$	$t = 14$
$R_1 = 47.5°$	$S_1 = 97.5°$	$T = 35°$

or

$r = 18$	$s_2 = 5.29$	$t = 14$
$R_2 = 132.5°$	$S_2 = 12.5°$	$T = 35°$

10. (a) Solve for R and Q using the Law of Cosines. Solve for P using $P + Q + R = 180°$.

$p = 300$	$q = 271$	$r = 415$
$P = 46.2°$	$Q = 40.7°$	$R = 93.1°$

11. (a) Solve for D using the Law of Sines.
   (b) no solution

12. (a) Solve for x using the Law of Cosines. Solve for Y using the Law of Cosines. Solve for Z using $X + Y + Z = 180°$.

    (b) $x = 12.76$ $y = 7$ $z = 15$

         $X = 58°$ $Y = 27.7°$ $Z = 94.3°$

13. (a) Solve for R using $R + A = 90°$. Solve for a and r using trigonometric ratios.

    (b) $r = 9.74$ $a = 3.93$ $t = 10.5$

         $R = 68°$ $A = 22°$ $T = 90°$

14. (a) Solve for D and A using the Law of Cosines. Solve for M using $M + A + D = 180°$.

    (b) $m = 27$ $a = 16$ $d = 40$

         $M = 28.3°$ $A = 16.3°$ $D = 135.4°$

15. (a) Solve for Q using $P + D + Q = 180°$. Solve for d and p using the Law of Sines.

    (b) $p = 12.31$ $d = 15.4$ $q = 9.7$

         $P = 53°$ $D = 88°$ $Q = 39°$

16. (a) Solve for C using $B + C = 90°$. Solve for a and b using trigonometric ratios.

    (b) $a = 182.26$ $b = 177.59$ $c = 41$

         $A = 90°$ $B = 77°$ $C = 13°$

17. (a) Solve for W using the Law of Sines. Solve for A using $T + W + A = 180°$. Solve for a using the Law of Sines.

    (b) $t = 52$ $w = 47$ $a = 96.59$

         $T = 13°$ $W = 11.7°$ $A = 155.3°$

18. (a) Solve for m using the Law of Cosines. Solve for P using the Law of Cosines. Solve for N using $M + N + P = 180°$.

    (b) $m = 24.31$ $n = 8$ $p = 31$

         $M = 29°$ $N = 9.2°$ $P = 141.8°$

Objective 4.3

1. 686.87 meters

2. 1032.35 feet

3. (a) 2.44 miles

    (b) 2.81 miles

    (c) 1.69 miles

4. 51.3°

5. 44.7°

6. 113.29 feet

7. 395.96 square inches

8. (a) 128.56 miles

    (b) 153.21 miles

9. 97.93 feet

10. 64.8 feet

11. 295.75 feet or 108.97 feet

12. 3.2 meters

13. 13923 meters

14. 29.86 feet

15. 46.44 miles

16. 83.13 meters

17. (a) N46.7°W

    (b) S36.5°W

18. 323.6 feet

19. (a) 68°

    (b) 30.2°

20. 81.68 miles

21. 46.6°

22. 70.46 meters

23. 46.1°

24. (a) 14.11 feet

    (b) 19.42 feet

25. 77.72 feet

26. 4.98 inches and 6.42 inches

27. 42.81 miles

28. 1039.23 feet

29. 40.47 square feet

30. 84.7 miles

Sample Examination 1

1. Sketch and label the triangle. By the Law of Cosines,
$$(18.6)^2 = (24.5)^2 + (26.4)^2 - 2(24.5)(26.4) \cos B.$$

Therefore, $\cos B = \dfrac{(24.5)^2 + (26.4)^2 - (18.6)^2}{2(24.5)(26.4)} = 0.73535$ and $B = 42.7°$.

2. Sketch and label the triangle. By the Law of Cosines,
$$c^2 = (132)^2 + (224)^2 - 2(132)(224) \cos 28° = 15386.011 \text{ and } c = 124.$$

Perimeter $= 132 + 224 + 124 = 480$.

3. Sketch and label the triangle. First, $C = 180° - 75° - 38° = 67°$. Then, by the Law of Sines,
$$\frac{b}{\sin 38°} = \frac{321}{\sin 67°} \quad \text{and} \quad b = \frac{321 \sin 38°}{\sin 67°} = 214.7.$$

4. Draw a sketch and label the known quantities. Label the distance from the top of the crate to the dock d as in **4.4A**. This is a right triangle, so

$$\sin 75° = \frac{10 + d}{80} \quad \text{and} \quad 10 + d = 80 \sin 75° = 77.27.$$

Therefore, $d = 77.27 - 10 = 67.27$ feet.

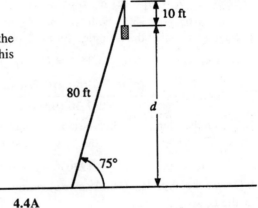

4.4A

5. Draw a sketch. Label the beginning of the path A, the rifleman B and the point where the target is first within range of the rifle C as in **4.5A**. Distance b is to be found. This is the ambiguous case of the Law of Sines. We have

$$\frac{125}{\sin C} = \frac{100}{\sin 40°} \quad \text{and}$$

$$\sin C = \frac{125 \sin 40°}{100} = 0.80348.$$

The possible solutions are

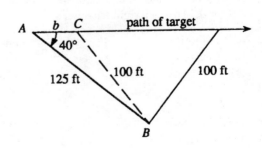

4.5A

$$C_1 = 53.5° \quad \text{and} \quad C_2 = 126.5°.$$

Since $C_1 + 40° = 93.5° < 180°$ and $C_2 + 40° = 166.5° < 180°$, both C_1 and C_2 are solutions. From **4.5A** we see that $C = C_2 = 126.5°$ is the desired angle. Solve triangle ABC for side b. Now $B = 180° - 40° - 126.5° = 13.5°$, so, by the Law of Sines,

$$\frac{b}{\sin 13.5°} = \frac{100}{\sin 40°} \quad \text{and} \quad b = \frac{100 \sin 13.5°}{\sin 40°} = 36.3 \text{ feet.}$$

Sample Examination 2

1. Sketch and label the triangle. By the Law of Sines,

$$\frac{\sin A}{62.5} = \frac{\sin 40°}{51.5} \quad \text{and} \quad \sin A = \frac{62.5 \sin 40°}{51.5} = 0.78008.$$

Therefore, $A = A_1 = 51.3°$ or $A = A_2 = 180° - A_1 = 128.7°$. Since $A_1 + B = 91.3° < 180°$ and $A_2 + B = 168.7° < 180°$, both A_1 and A_2 are solutions.

2. Sketch and label the triangle. By the Law of Cosines,

$$(6.34)^2 = (7.30)^2 + (9.98)^2 - 2(7.30)(9.98) \cos A.$$

Therefore, $\cos A = \dfrac{(7.30)^2 + (9.98)^2 - (6.34)^2}{2(7.30)(9.98)} = 0.77339$ and $A = 39.3°$.

3. Sketch the triangle. Label the side having length 120 a, the side having length 270 b, and the remaining side c. Then $C = 118°$. By the Law of Cosines,

$$c^2 = (120)^2 + (270)^2 - 2(120)(270) \cos 118° = 117721.76 \text{ and } c = 343.1.$$

4. Draw a sketch and label the known quantities. Label the distance from the ball to the hole c. Then $C = 10°$. By the Law of Cosines,

$$c^2 = (205)^2 + (340)^2 - 2(205)(340) \cos 10° = 20342.8 \text{ and } c = 142.6 \text{ yards.}$$

5. Draw a sketch and label the known quantities. Label the distance between the ridge of the roof and the attic floor d. This is a right triangle, so

$$\sin 18° = \frac{d}{12.94} \quad \text{and} \quad d = 12.94 \sin 18° = 3.99 \text{ feet.}$$

UNIT 5
GRAPHING TRIGONOMETRIC FUNCTIONS

Objective 5.1

1. $\dfrac{\sqrt{3}}{3}$

2. $\sqrt{2}$

3. undefined

4. $2 + \sqrt{3}$

5. $\dfrac{\sqrt{2}}{2}$

6. $\dfrac{1}{4}$

7. $\dfrac{-\sqrt{3}}{3}$

8. undefined

9. $\dfrac{\sqrt{2}}{4}\left(\sqrt{3}-1\right)$

10. 0

11. 1

12. $\dfrac{1}{2}$

13. 7

14. $\dfrac{2}{3}\left(3-\sqrt{3}\right)$

15. $\sqrt{2}$

16. $\dfrac{8}{3}$

17. $2\sqrt{2}$

18. $\dfrac{-2}{3}$

19. (E)

20. (B)

Objective 5.2

1. 1   **2.** 2π **3.** 2π **4.** 1

5.

θ	$\sin\theta$	$\cos\theta$
0	0	1
$\dfrac{\pi}{12}$	0.26	0.97
$\dfrac{\pi}{6}$	0.5	0.87
$\dfrac{\pi}{4}$	0.71	0.71
$\dfrac{\pi}{3}$	0.87	0.5
$\dfrac{5\pi}{12}$	0.97	0.26

θ	$\sin\theta$	$\cos\theta$
$\dfrac{\pi}{2}$	1	0
$\dfrac{7\pi}{12}$	0.97	-0.26
$\dfrac{2\pi}{3}$	0.87	-0.5
$\dfrac{3\pi}{4}$	0.71	-0.71
$\dfrac{5\pi}{6}$	0.5	-0.87
$\dfrac{11\pi}{12}$	0.26	-0.97

θ	$\sin\theta$	$\cos\theta$
π	0	-1
$\dfrac{13\pi}{12}$	-0.26	-0.97
$\dfrac{7\pi}{6}$	-0.5	-0.87
$\dfrac{5\pi}{4}$	-0.71	-0.71
$\dfrac{4\pi}{3}$	-0.87	-0.5
$\dfrac{17\pi}{12}$	-0.97	-0.26

θ	$\sin \theta$	$\cos \theta$
$\dfrac{3\pi}{2}$	-1	0
$\dfrac{19\pi}{12}$	-0.97	0.26
$\dfrac{5\pi}{3}$	-0.87	0.5
$\dfrac{7\pi}{4}$	-0.71	0.71
$\dfrac{11\pi}{6}$	-0.5	0.87
$\dfrac{23\pi}{12}$	-0.26	0.97
2π	0	1

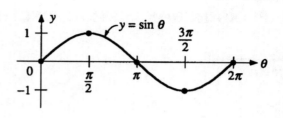

6. 5.9A

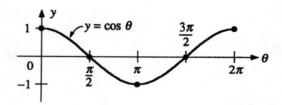

7. 5.10A

8. $A = (-\pi, 0)$ $B = \left(\dfrac{-\pi}{2}, -1\right)$ $C = \left(\dfrac{\pi}{2}, 1\right)$ $D = (\pi, 0)$

     $E = (2\pi, 0)$ $F = \left(\dfrac{5\pi}{2}, 1\right)$ $G = \left(\dfrac{7\pi}{2}, -1\right)$

9. $A = (-2\pi, 1)$ $B = (-\pi, -1)$ $C = \left(\dfrac{-\pi}{2}, 0\right)$ $D = (\pi, -1)$

     $E = (2\pi, 1)$ $F = \left(\dfrac{5\pi}{2}, 0\right)$ $G = \left(\dfrac{7\pi}{2}, 0\right)$

10. (C)   **11. (B)**

Objective 5.3

1. amplitude $= 5$ period $= 2\pi$

2. amplitude $= \dfrac{1}{3}$ period $= 2\pi$

3. amplitude $= 1$ period $= 8\pi$

4. amplitude $= 1$ period $= \pi$

5. amplitude $= 2$ period $= \dfrac{2\pi}{3}$

6. amplitude $= 2$ period $= 6\pi$

7. amplitude $= \dfrac{3}{2}$ period $= \pi$

8. amplitude $= \dfrac{7}{3}$ period $= 4\pi$

9. amplitude $= 3$ period $= 2$

10. amplitude $= 7$ period $= 1$

11. amplitude $= \dfrac{5}{4}$ period $= 4$

12. amplitude $= \dfrac{3}{2}$ period $= 3$

13. amplitude $= 2$ period $= \dfrac{2\pi}{3}$, $y = 2\cos 3\theta$

14. amplitude $= \dfrac{3}{2}$ period $= 4\pi$, $y = \dfrac{3}{2}\sin\dfrac{1}{2}\theta$

15. $y = 3\cos\dfrac{4}{3}\theta$

16. $y = \dfrac{1}{2}\sin\dfrac{3}{5}\theta$

17. $y = \dfrac{2}{3}\cos \pi\theta$

18. $y = \dfrac{9}{4}\sin\dfrac{3\pi}{2}\theta$

Objective 5.4

1. 2π **3.** 2π **5.** 2π

2. π **4.** π **6.** 2π

7.

θ	$\cot \theta$
0	undefined
$\dfrac{\pi}{12}$	3.73
$\dfrac{\pi}{6}$	1.74
$\dfrac{\pi}{4}$	1
$\dfrac{\pi}{3}$	0.57
$\dfrac{5\pi}{12}$	0.27

θ	$\cot \theta$
$\dfrac{\pi}{2}$	0
$\dfrac{7\pi}{12}$	-0.27
$\dfrac{2\pi}{3}$	-0.57
$\dfrac{3\pi}{4}$	-1
$\dfrac{5\pi}{6}$	-1.74
$\dfrac{11\pi}{12}$	-3.73
π	undefined

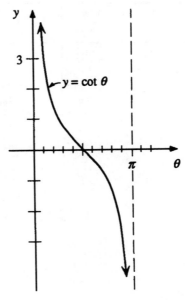

$y = \cot \theta$

5.32A

204

8.

θ	$\csc \theta$	θ	$\csc \theta$
0	undefined	π	undefined
$\dfrac{\pi}{12}$	3.86	$\dfrac{13\pi}{12}$	−3.86
$\dfrac{\pi}{6}$	2	$\dfrac{7\pi}{6}$	−2
$\dfrac{\pi}{4}$	1.41	$\dfrac{5\pi}{4}$	−1.41
$\dfrac{\pi}{3}$	1.15	$\dfrac{4\pi}{3}$	−1.15
$\dfrac{5\pi}{12}$	1.04	$\dfrac{17\pi}{12}$	−1.04
$\dfrac{\pi}{2}$	1	$\dfrac{3\pi}{2}$	−1
$\dfrac{7\pi}{12}$	1.04	$\dfrac{19\pi}{12}$	−1.04
$\dfrac{2\pi}{3}$	1.15	$\dfrac{5\pi}{3}$	−1.15
$\dfrac{3\pi}{4}$	1.41	$\dfrac{7\pi}{4}$	−1.41
$\dfrac{5\pi}{6}$	2	$\dfrac{11\pi}{6}$	−2
$\dfrac{11\pi}{12}$	3.86	$\dfrac{23\pi}{12}$	−3.86
		2π	undefined

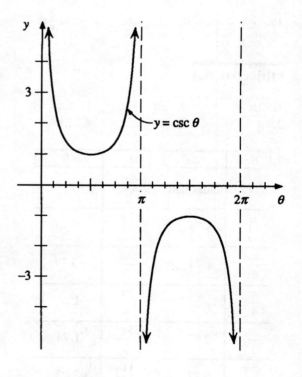

$y = \csc \theta$

5.33A

9.

θ	$\sec \theta$	θ	$\sec \theta$
$\dfrac{-\pi}{2}$	undefined	$\dfrac{\pi}{2}$	undefined
$\dfrac{-5\pi}{12}$	3.86	$\dfrac{7\pi}{12}$	−3.86
$\dfrac{-\pi}{3}$	2	$\dfrac{2\pi}{3}$	−2
$\dfrac{-\pi}{4}$	1.53	$\dfrac{3\pi}{4}$	−1.53
$\dfrac{-\pi}{6}$	1.15	$\dfrac{5\pi}{6}$	−1.15
$\dfrac{-\pi}{12}$	1.04	$\dfrac{11\pi}{12}$	−1.04
0	1	π	−1
$\dfrac{\pi}{12}$	1.04	$\dfrac{13\pi}{12}$	−1.04
$\dfrac{\pi}{6}$	1.15	$\dfrac{7\pi}{6}$	−1.15
$\dfrac{\pi}{4}$	1.53	$\dfrac{5\pi}{4}$	−1.53
$\dfrac{\pi}{3}$	2	$\dfrac{4\pi}{3}$	−2
$\dfrac{5\pi}{12}$	3.86	$\dfrac{17\pi}{12}$	−3.86
		$\dfrac{3\pi}{2}$	undefined

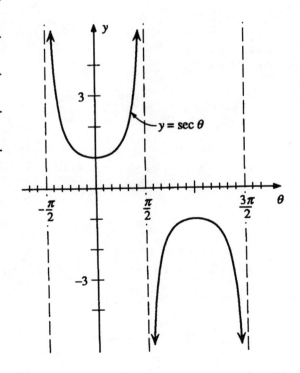

$y = \sec \theta$

5.34A

Sample Examination 1

1. $\sin\dfrac{\pi}{4}\cot\dfrac{\pi}{6} - \csc 60° \tan 45° = \dfrac{\sin \pi/4}{\tan \pi/6} - \dfrac{\tan 45°}{\sin 60°} = \dfrac{\sqrt{2}/2}{1/\sqrt{3}} - \dfrac{1}{\sqrt{3}/2} = \dfrac{\sqrt{6}}{2} - \dfrac{2}{\sqrt{3}} = \dfrac{3\sqrt{6} - 4\sqrt{3}}{6}$.

2. One cycle of the sine curve occurs over the interval from $x = 0$ to $x = 2\pi$. The point P is three-fourths

of the way through the cycle, so the x-coordinate of P is $x = \dfrac{3}{4}(2\pi) = \dfrac{3\pi}{2}$. The minimum value of the sine

function is $y = -1$. The coordinates of the point P are $\left(\dfrac{3\pi}{2}, -1\right)$.

3. The period of $y = \dfrac{2}{3}\cos\dfrac{\pi}{2}\theta$ is $P = \dfrac{2\pi}{\pi/2} = 4$.

4. Since y is at a maximum when $\theta = 0$, this is a graph of a function of the form $y = A\cos B\theta$. Since

the graph oscillates 2 units above and below the horizontal axis, the amplitude is 2 and $A = 2$. The interval

from 0 to $\dfrac{4}{\pi}$ contains exactly one cycle of the graph so the period is $\dfrac{4}{\pi}$. This means that $\dfrac{2\pi}{B} = \dfrac{4}{\pi}$ and

$B = \dfrac{\pi^2}{2}$. The equation for the function is $y = 2\cos\dfrac{\pi^2}{2}\theta$.

5. This is the graph of $y = -\sec\theta$.

Sample Examination 2

1. $\csc\dfrac{\pi}{6}\cos\dfrac{\pi}{3} + \sin 45° \cot 30° = \dfrac{\cos \pi/3}{\sin \pi/6} + \dfrac{\sin 45°}{\tan 30°} = \dfrac{1/2}{1/2} + \dfrac{\sqrt{2}/2}{1/\sqrt{3}} = 1 + \dfrac{\sqrt{6}}{2} = \dfrac{2 + \sqrt{6}}{2}$.

2. One cycle of the cosine curve occurs over the interval from $x = 0$ to $x = 2\pi$. The point Q is at the
end of the cycle, so the x-coordinate of Q is $x = 2\pi$. The maximum value of the cosine function is $y = 1$.
The coordinates of the point Q are $\left(2\pi, 1\right)$.

3. The amplitude of $y = 5\sin\dfrac{1}{3}\theta$ is $A = 5$.

4. Since $y = 0$ when $\theta = 0$, this is a graph of a function of the form $y = A\sin B\theta$. Since the graph
oscillates three units above and below the horizontal axis, the amplitude is 3 and $A = 3$. The interval from

0 to $\dfrac{4\pi}{3}$ contains exactly one cycle of the graph so the period is $\dfrac{4\pi}{3}$. This means that $\dfrac{2\pi}{B} = \dfrac{4\pi}{3}$ and $B = \dfrac{3}{2}$.

The equation for the function is $y = 3\sin\dfrac{3}{2}\theta$.

5. This is the graph of $y = \cot\theta$.

UNIT 6
COURSE REVIEW

Objective 6.1
Sample Examination

1. Since $1° = \dfrac{\pi}{180}$ radians, $972° = 972\left(\dfrac{\pi}{180}\right)$ radians $= 16.96$ radians.

2. Angle θ was formed by rotating the ray clockwise through $\dfrac{7}{8}$ revolution, so $\theta = \left(-\dfrac{7}{8}\right)(360°) = -315°$.

3. Since the point $(-2, 3)$ is on the terminal side of θ, choose $x = -2$, $y = 3$, $r = \sqrt{4+9} = \sqrt{13}$. Then,

$\sin \theta = \dfrac{y}{r} = \dfrac{3}{\sqrt{13}}$, $\sec \theta = \dfrac{r}{x} = -\dfrac{\sqrt{13}}{2}$ and, rounded to two decimal places,

$$3 \sin \theta + 7 \sec \theta = 3\left(\dfrac{3}{\sqrt{13}}\right) + 7\left(-\dfrac{\sqrt{13}}{2}\right) = -10.12.$$

4. Since $\cos \theta < 0$ and $\tan \theta > 0$, θ is a quadrant III angle. Choose (x, y) on the terminal side of θ

so $x < 0$, $y < 0$, and $\cos \theta = \dfrac{x}{r} = \dfrac{-5}{6}$. The easiest choice is $r = 6$, $x = -5$ and

$y = -\sqrt{(6)^2 - (5)^2} = -\sqrt{11}$. Then, $\csc \theta = \dfrac{r}{y} = -\dfrac{6}{\sqrt{11}}$, $\cot \theta = \dfrac{x}{y} = \dfrac{-5}{-\sqrt{11}} = \dfrac{5}{\sqrt{11}}$ and, rounded to

three decimal places,

$$4 \csc \theta - \cot \theta = 4\left(-\dfrac{6}{\sqrt{11}}\right) - \dfrac{5}{\sqrt{11}} = -8.744.$$

5. Set radian mode. Rounded to three decimal places,

$$7 \csc (7.36) \cot \left(-\dfrac{3\pi}{8}\right) - 2 \cos (2.78) = 7(1.1358)(-0.4142) - 2(-0.9393) = -1.423.$$

6. Set degree mode. Rounded to four decimal places,

$5 \csc (212°) \cot (750°) + 9 \cos (-438°) = 5(-1.887080)(1.732051) + 9(0.207912) = -14.4714.$

7. Set degree mode. Since $\sin S = -0.4387$, S is a quadrant III or quadrant IV angle with reference angle $S' = 26°$. The possible values of S are $S_1 = 206°$ and $S_2 = 334°$. Since $\tan T = -6.8974$, T is a quadrant II or quadrant IV angle with reference angle $T' = 82°$. The possible values of T are $T_1 = 98°$ and $T_2 = 278°$. The possible values of $S + T$ are

$S_1 + T_1 = 304°$, $\qquad$ $S_2 + T_1 = 432°$, $\qquad$ $S_1 + T_2 = 484°$, $\qquad$ $S_2 + T_2 = 612°$.

8. Set radian mode. Since $\cos \alpha = 0.4599$, α is a quadrant I or quadrant IV angle with reference angle $\alpha' = 1.0929$. The possible values of α are $\alpha_1 = 1.0929$ and $\alpha_2 = 5.1903$. Since $\sin \beta = -0.8797$, β is a quadrant III or quadrant IV angle with reference angle $\beta' = 1.0752$. The possible values of β are $\beta_1 = 4.2168$ and $\beta_2 = 5.2080$. The possible values of $\alpha + \beta$ are

$$\alpha_1 + \beta_1 = 5.310 \qquad \alpha_2 + \beta_1 = 9.397 \qquad \alpha_1 + \beta_2 = 6.301 \qquad \alpha_2 + \beta_2 = 10.398.$$

9. Sketch and label the triangle. This is a right triangle. Thus,

$$\sin 17° = \frac{2.32}{b} \quad \text{and} \quad b = \frac{2.32}{\sin 17°} = 7.94$$

$$\tan 17° = \frac{2.32}{c} \quad \text{and} \quad c = \frac{2.32}{\tan 17°} = 7.59.$$

Perimeter $= 2.32 + 7.94 + 7.59 = 17.85$.

10. Sketch and label the triangle. This is a right triangle. Therefore,

$$p^2 = q^2 - r^2 = (12.79)^2 - (3.89)^2 = 148.45 \text{ and } p = 12.18.$$

11. Sketch and label the triangle. Compute $A = 180° - B - C = 22°$. By the Law of Sines,

$$\frac{c}{\sin 101°} = \frac{16.3}{\sin 22°} \quad \text{and} \quad c = \frac{16.3 \sin 101°}{\sin 22°} = 42.7.$$

12. Sketch and label the triangle. By the Law of Cosines,

$$(33.3)^2 = (25.1)^2 + (50.4)^2 - 2(25.1)(50.4) \cos C.$$

Therefore,

$$\cos C = \frac{(25.1)^2 + (50.4)^2 - (33.3)^2}{2(25.1)(50.4)} = 0.8147 \text{ and } C = 35.4°.$$

13. Draw a sketch and label the known quantities. Label the length of the third side l. This is a right triangle

so $l = \sqrt{(521.5)^2 + (636.4)^2} = 822.8$ feet.

14. Draw a sketch. Label the top if the stairway T and the length of the brass rail l.

Angle $T = 90° - 34° = 56°$. Since this is a right triangle

$$\cos 56° = \frac{2.5}{l}, \quad l = \frac{2.5}{\cos 56°} = 4.47 \text{ meters}.$$

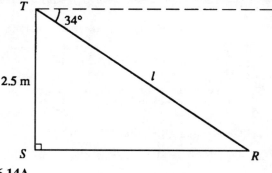

6.14A

15. Draw a sketch. Label the two sides of length 12 centimeters a and b and the distance between the tips c. Then $C = 22°$. By the Law of Cosines,

$$c^2 = (12)^2 + (12)^2 - 2(12)(12) \cos 22° = 20.97 \text{ and } c = 4.6 \text{ centimeters}.$$

16. Draw a sketch. Label the window W, the base of the shorter ladder S and the base of the longer ladder L. Label the known quantities. First, find angle S of triangle WSL.

$$S = 180° - 75° = 105°$$

By the Law of Sines,

$$\frac{\sin 105°}{24} = \frac{\sin L}{20} \quad \text{and} \quad \sin L = \frac{20 \sin 105°}{24} = 0.8049.$$

The possible values for L are $L_1 = 54°$ and $L_2 = 126°$. Since we are looking for the acute angle, $L = L_1 = 54°$.

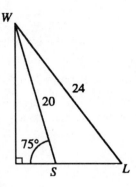

6.16A

17. $\csc \dfrac{\pi}{6} \tan \dfrac{\pi}{3} - \cot 45° \sin 60° = \dfrac{\tan \frac{\pi}{3}}{\sin \frac{\pi}{6}} - \dfrac{\sin 60°}{\tan 45°} = \dfrac{\sqrt{3}}{\frac{1}{2}} - \dfrac{\frac{\sqrt{3}}{2}}{1} = \dfrac{3\sqrt{3}}{2}.$

18. Since y is at a maximum when $\theta = 0$, this is a graph of a function of the form $y = A \cos B\theta$. Since the graph oscillates $\dfrac{2}{5}$ units above and below the horizontal axis, the amplitude is $\dfrac{2}{5}$ and $A = \dfrac{2}{5}$. The interval from 0 to 4 contains exactly one cycle of the graph so the period is 4. This means that

$$\frac{2\pi}{B} = 4 \quad \text{and} \quad B = \frac{\pi}{2}.$$

The equation for the function is $y = \dfrac{2}{5} \cos \dfrac{\pi}{2}.$

19. Since y is at a maximum when $\theta = 0$, this is a graph of a function of the form $A \cos B\theta$. Since the graph oscillates π units above and below the horizontal axis, the amplitude is π and $A = \pi$. Since the interval from 0 to $\dfrac{3\pi}{2}$ contains exactly one cycle of the graph, the period is $\dfrac{3\pi}{2}$. This means that

$$\frac{2\pi}{B} = \frac{3\pi}{2} \quad \text{and} \quad B = \frac{4}{3}.$$

The equation for the function is $y = \pi \cos \dfrac{4}{3}\theta.$

20. This is the graph of $y = \csc \theta$.

INDEX

Index